RECENT RESEARCH IN POLYMERIZATION

Edited by **Nevin Çankaya**

Recent Research in Polymerization
http://dx.doi.org/10.5772/intechopen.68265
Edited by Nevin Çankaya

Contributors

Alex A. Lyapkov, Stanislav Kiselev, Galina Bozhenkova, Mekhman Yusubov, Olga Kukurina, Francis Verpoort, Abdelaziz El-Hoshoudy, Gaelle Aziz, Rouba Ghobeira, Nathalie De Geyter, Rino Morent, Bianca Rita Pistillo, Damien Lenoble, Kevin Menguelti, Didier Arl, Renaud Leturcq, Mihrigul Altan, Roberto Fernández, Victor Navarro Fuster, Francisco Javier Martínez, Sergi Gallego, Andrés Márquez, Cristian Neipp López, Inmaculada Pascual, Augusto Beléndez, Lawrence Madikizela, Nikita Tavengwa, Vusumzi Emmanuel Pakade, Igor Denisyuk, Mayya Uspenskaya, Roman Olekhnovich, Michael Dovbeta, Vera Sitnikova, Yanjie Wang, Denglin Zhu, Minzhou Luo, Xiaofeng Liu, Hualing Chen, Jie Ru

Published by InTech
Janeza Trdine 9, 51000 Rijeka, Croatia

Publishing Process Manager Dajana Pemac
Technical Editor SPi Global
Cover InTech Design team

First published January, 2018
Printed in Croatia
Legal deposit, Croatia: National and University Library in Zagreb

Additional hard copies can be obtained from orders@intechopen.com

Recent Research in Polymerization, Edited by Nevin Çankaya
p. cm.
Print ISBN 978-953-51-3746-7
Online ISBN 978-953-51-3747-4

For EU product safety concerns:
IN TECH d.o.o., Prolaz Marije Krucifikse Kozulić 3, 51000 Rijeka, Croatia,
info@intechopen.com or visit our website at intechopen.com.

Meet the editor

Nevin Çankaya is associated with the Department of Chemistry, Faculty of Science, Uşak University, since 2011. She has received her PhD, master's, and bachelor's degrees from Fırat University, Turkey. She is an expert in polymer composite science, working with bionano-composites, polymeric materials, and polymer-clay composites. She has published over 20 scientific articles and over 30 conference papers and has been a manager and researcher in many research projects. Moreover, she is the editor of *Elastomers* (ISBN 978-953-51-3488-6) published by InTechOpen, as well as a coauthor of *Cellulose-Fundamental Aspects and Current Trends* (ISBN 978-953-51-2229-6). She is married with two children.

Contents

Preface VII

Section 1 General Information About Polymerization 1

Chapter 1 **Emulsion Polymerization Mechanism 3**
Abdelaziz Nasr Moawed Bakr El-hoshoudy

Chapter 2 **Ring Opening Metathesis Polymerization 15**
Alexey Lyapkov, Stanislav Kiselev, Galina Bozhenkova, Olga Kukurina, Mekhman Yusubov and Francis Verpoort

Section 2 Biomaterial Content Polymer Composites 45

Chapter 3 **Molecularly Imprinted Polymers for Pharmaceutical Compounds: Synthetic Procedures and Analytical Applications 47**
Lawrence Madikizela, Nikita Tavengwa and Vusumzi Pakade

Chapter 4 **Plasma Polymerization for Tissue Engineering Purposes 69**
Gaelle Aziz, Rouba Ghobeira, Rino Morent and Nathalie De Geyter

Chapter 5 **Sorption Properties of Clay and Pectin-Containing Hydrogels 95**
Mayya V. Uspenskaya, Vera E. Sitnikova, Michael A. Dovbeta, Roman O. Olekhnovich and Igor Yu. Denisyuk

Section 3 Mechanical Properties of Polymerization 115

Chapter 6 **Thermoplastic Foams: Processing, Manufacturing, and Characterization 117**
Mihrigul Altan

Chapter 7 **PRAP-CVD: A Novel Technique to Deposit Intrinsically Conductive Polymers 139**
Bianca Rita Pistillo, Kevin Menguelti, Didier Arl, Renaud Leturcq and Damien Lenoble

Chapter 8 **Polymerizable Materials for Diffractive Optical Elements Recording 161**
Roberto Fernández Fernández, Víctor Navarro Fuster, Francisco J. Martínez Guardiola, Sergi Gallego Rico, Andrés Márquez Ruiz, Cristian Neipp López, Inmaculada Pascual Villalobos and Augusto Beléndez Vázquez

Chapter 9 **The Effects of Casting and Blending on Properties of Ionomer and the Electromechanical Responses of Ionic Polymer Metal Composite Actuators 187**
Yanjie Wang, Jie Ru, Denglin Zhu, Minzhou Luo, Xiaofeng Liu and Hualing Chen

Preface

The purpose of this book is to help you concentrate on recent developments in polymerization. The chapters collected in the book are contributions by invited researchers with a long-standing experience in different research areas. I hope that the material presented here is understandable to a broad audience, not only chemists but also scientists from various disciplines. The book contains nine chapters in three sections: (1) "General Information about Polymerization," (2) "Biomaterial Content Polymer Composites," and (3) "Mechanical Properties of Polymerization." The book provides detailed and current reviews in these different areas written by experts in their respective fields. This book is expected to be useful for polymer workers and other scientists alike and contribute to the training of current and future researchers, academics, PhD degree students, as well as other scientists.

Asst. Prof. Nevin Çankaya
Uşak University,
Turkey

Section 1

General Information About Polymerization

Chapter 1

Emulsion Polymerization Mechanism

Abdelaziz Nasr Moawed Bakr El-hoshoudy

Additional information is available at the end of the chapter

http://dx.doi.org/10.5772/intechopen.72143

Abstract

Emulsion polymerization is a polymerization process with different applications on the industrial and academic scale. It involves application of emulsifier to emulsify hydrophobic polymers through aqueous phase by amphipathic emulsifier, then generation of free radicals with either a water or oil soluble initiators. It characterized by reduction of bimolecular termination of free radicals due to segregation of free radicals among the discrete monomer-swollen polymer particles. The latex particles size ranged from 10 nm to 1000 nm in a diameter and are generally spherical. A typical of particle consist of 1–10,000 macromolecules, where macromolecule contains about 100–10^6 monomer units.

Keywords: emulsion polymerization, emulsified monomers, particle nucleation and polymerization mechanism

1. Introduction

Emulsion polymerization is a unique process involves emulsification of hydrophobic monomers by oil-in water emulsifier, then reaction initiation with either a water soluble initiator (e.g. potassium persulfate ($K_2S_2O_8$) or an oil-soluble initiator (e.g. 2,2-azobisisobutyronitrile (AIBN)) [1, 2] in the presence of stabilizer which may be ionic, nonionic or protective colloid to disperse hydrophobic monomer through aqueous solution [3, 4]. Typical polymerization monomers involve vinyl monomers of the structure (CH_2=CH-). These emulsion polymers find a wide range of applications such as synthetic rubbers, thermoplastics, coatings, adhesives, binders, rheological modifiers, plastic pigments [1]. Emulsion polymerization is a rather complex process because nucleation, growth and stabilization of polymer particles are controlled by the free radical polymerization mechanisms in combination with various colloidal phenomena [1]. Aside from other polymerization techniques, emulsion polymerization affords increasing molecular weight of the formed latexes through decreasing polymerization rate by either decreasing initiator concentration or lowering reaction temperature [5, 6].

Systems of emulsion polymerization involve (1) conventional emulsion polymerization, in which a hydrophobic monomer emulsified in water and polymerization initiated with a water-soluble initiator [5]. (2) Inverse emulsion polymerization [7], where organic solvents of very low polarity as paraffin or xylene used as a polymerization media to emulsify hydrophilic monomers [5], then initiation proceed with the aid of hydrophobic initiator [5]. These two polymerization types known as oil-in-water (o/w) and water-in-oil (w/o) emulsions [5]. (3) Mini emulsion polymerization involves systems with monomer droplets in water with much smaller droplets than in emulsion polymerization and characterized by monomer droplet =50–1000 nm, surfactant concentration < critical micelle concentration (CMC), water insoluble co stabilizer as hexadecane to prevent Ostwald ripening, polymer particle size equal monomer droplet size = 50–1000 nm, and both water soluble and oil soluble initiator used [4, 8]. (4) Microemulsion polymerization with very much smaller monomer droplets, about 10–100 nm, and characterized by surfactant concentration > CMC, polymer particles = 10–50 nm, water-soluble initiator are commonly used [9, 10]. Miniemulsion, microemulsion and conventional emulsion polymerizations show quite different particle nucleation and growth mechanisms and kinetics [1]. Many articles discuss different types of emulsion polymerization found in literature [1, 11–16].

2. Components of heterogeneous emulsion polymerization

The main components of emulsion polymerization media involve monomer(s), dispersing medium, emulsifier, and water-soluble initiator [5, 17–19]. The dispersion medium is water in which hydrophobic monomers is emulsified by surface-active agents (surfactant). When surfactant concentration exceeds critical micelle concentration (CMC) it aggregate in the form of spherical micelles, so surface tension at the surface decrease, as a result hydrophobic monomers enter in to the vicinity of micelle and reaction continue until all monomer droplets are exhausted and micelle containing monomers increase in size. Typical micelles have dimensions of 2–10 nm, with each micelle containing 50–150 surfactant molecules [5]. Water-soluble initiators enter into the micelle where free radical propagation start. In general, monomer droplets are not effective in competing with micelles in capturing free radicals generated in the aqueous phase due to their relatively small surface area [1], so the micelle act as a meeting site of water-soluble initiators and hydrophobic vinyl monomers. As polymerization continue inside micelle, the micelle grow by monomer addition from monomer droplets outside and latex are formed. Schematic representation of emulsion polymerization shown in **Figure 1**. Emulsion polymerization carried out through three main intervals as shown in **Figure 2**.

There is a separate monomer phase in intervals I. The particle number increases with time in interval I and particle nucleation occurs in interval I. At the end of this stage most of surfactants are exhausted (i.e. micelles are exhausted) [5]. About one of every 102–103 micelles can be successfully converted into latex particles [1]. Particle nucleation process is greatly affected by surfactant concentration, which in turn affect particle size and particle size distribution of latex [1]. The lower the surfactant concentration, the lower the nucleation period the narrow the particle size distribution. At interval II (Particle growth stage), the polymerization continue and polymer particles increase in size until monomer droplets exhausted. Monomer droplets act as reservoirs to supply the growing particles with monomer and surfactant species. At interval

http://dx.doi.org/10.5772/intechopen.72143

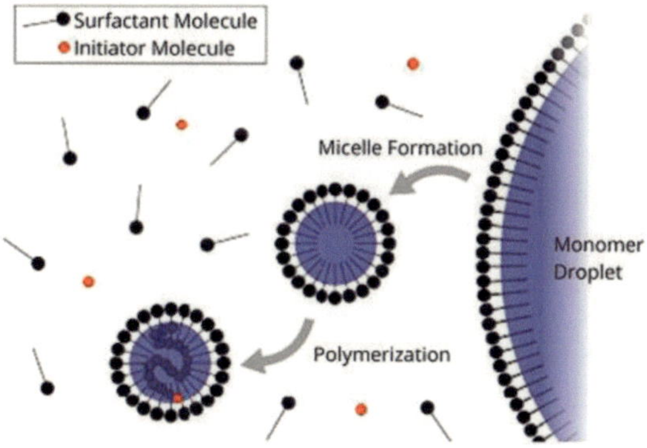

Figure 1. Schematic representation of emulsion polymerization.

III, the polymer size increase as latex particles become monomer-starved and the concentration of monomer in the reaction loci continues to decrease toward the end of polymerization [1].

2.1. Initiators

Initiator act to generate free radicals by thermal decomposition, or redox reactions. The initiators may be; (1) water-soluble initiators like 2,2-Azobis(2-amidinopropane) dihydrochloride, $K_2S_2O_8$, APS (Ammonium persulfate) and (H_2O_2) hydrogen peroxide. (2) Partially water-soluble peroxides like t-butyl hydroperoxide and succinic acid peroxide and azo compounds such as 4,4-azobis(4-cyanopentanoic acid) [14]. (3) Redox systems such as persulfate with ferrous ion, cumyl hydroperoxide or hydrogen peroxide with ferrous, sulfite, or bisulfite ion [5, 20]. Other initiators such as surface active initiators which "inisurfs," for example; bis[2-(4′-sulfophenyl)alkyl]-2,2′-azodiisobutyrate ammonium salts

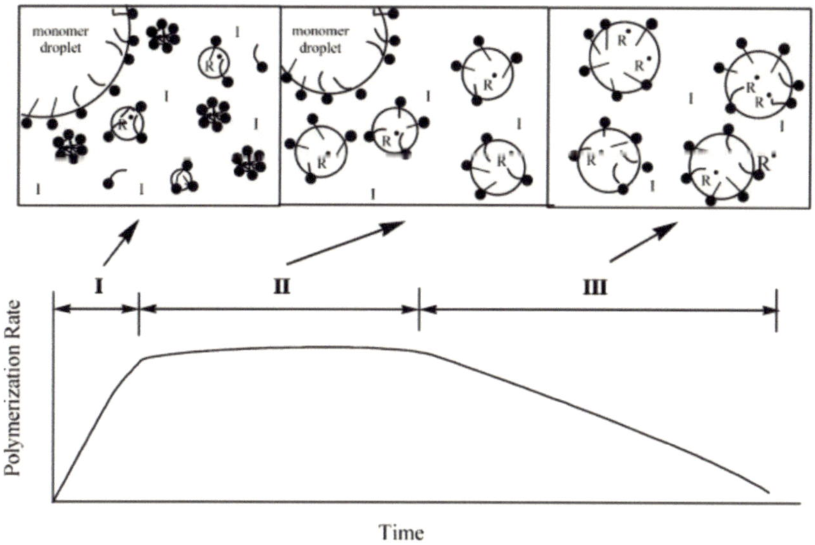

Figure 2. Emulsion polymerization intervals.

and 2,2′-azobis(N-2′-methylpropanoyl-2-amino-alkyl-1-sulfonate). These initiators initiates emulsion polymerization without the need of stabilizers.

2.2. Surfactants

Act to decrease interfacial tension between monomer and aqueous phase, stabilize the latex and generate micelles in which monomers emulsified and nucleation reaction proceed. Surfactants increase particle number and decrease particle size, these surfactants may be (1) Anionic surfactants such as fatty acid soaps (sodium or potassium stearate, laurate, palmitate), sulfates, and sulfonates (sodium lauryl sulfate and sodium dodecylbenzene sulfonate); (2) Nonionic surfactants such as poly (ethylene oxide), poly (vinyl alcohol) and hydroxyethyl cellulose; (3) Cationic surfactants such as dodecylammonium chloride and cetyltrimethylammonium bromide [5, 21]. For ionic surfactants, micelles formed only at temperatures above the Krafft point. For a nonionic surfactant, micelles formed only at temperatures below the cloud point. Emulsion polymerization carried out below the cloud temperature and above the Krafft temperature [5]. Polymerizable surfactants (surfactants with active double bond) such as sodium dodecyl allyl sulfosuccinate [13, 22–24] also used to produce latexes with chemically bound surface-active groups [5, 25–30, 31]. Polymerized surfactants (surfactants with active double bond) consist of amphipathic structure comprising hydrophobic tail and hydrophilic head group [32], in addition to polymerized vinyl groups [33] in their molecular structure, which acquire them unique physicochemical properties other than traditional surfactants moieties [34] such as;

A. They have surface activity like ordinary surfactants and polymerized vinyl group like vinyl monomers, so they have the ability to undergo polymerization reactions.

B. Due to their amphoteric structure and polymerization ability, they serve to synthesize inorganic/organic nanocomposite, and applied to emulsion polymerizations as polymerized emulsifiers, to stabilize the formed latexes, to prepare novel water-soluble hydrophobically associating polymers with strong thickening properties [35] so, they greatly applied in the field of enhanced oil recovery [36].

C. Allow developing hybrid Nano sized reaction and templating media. Moreover surfmer serve as hydrophilic monomer to copolymerize with acrylamide derivatives (AM) forming hydrophobically associating polyacrylamide (HAPAM), which acquire wide application in improved oil recovery coats and paintings and drilling fluids [37].

Freedman et al., [38] reported about the first synthesis of vinyl monomers which serve as emulsifying agents [39]. Active vinyl groups comprise vinyl, allyls, acrylates, methacrylates, styryl and acrylamide [40]. Polymerized groups may be "H-type" where, i.e. located in the hydrophilic head group, or "T-type" where, i.e. located in the hydrophobic tail have a profound effect on surfactant self-assembly and properties [41]. All kinds of polymerizable traditional surfactants, including cationic [41], anionic [42] and nonionic [43] have been synthesized to study the influence of the molecular structure on the properties and application. Anionic polymerizable surfactants seem to be the most promising for utilizing in coatings, adhesives and enhanced oil recovery.

http://dx.doi.org/10.5772/intechopen.72143

2.3. Dispersion medium

Water is the frequently used dispersion medium in emulsion polymerization as it is cheap and environmentally friendly. It represents the medium of transfer of monomer from droplets to particles and as a solvent for emulsifier, initiator, and other ingredients.

2.4. Monomer

Emulsion polymerization require free radical polymerizable monomers. Generally, vinyl monomers are used in this type of polymerization such as acrylamide, acrylic acid, butadiene, styrene, acrylonitrile, acrylate ester and methacrylate ester monomers, vinyl acetate, and vinyl chloride [1] and many other vinyl derivatives [25]. Depending on monomer solubility in aqueous phase, there is three categories of typical emulsion polymerization monomers which comprise (1) monomers of high solubility such as acrylonitrile, (2) monomers of medium solubility as methyl methacrylate and monomers insoluble in aqueous phase such as butadiene and styrene [44].

2.5. Other constituents

Other components involve emulsion polymerization medium that is generally deionized water. Antifreeze additives which involve inorganic electrolytes, ethylene glycol, glycerol, methanol, and monoalkyl ethers of ethylene glycol to allow polymerization at temperatures below 0°C. Sequestering agents which used to solubilize the initiator system or to deactivate traces of hardness elements (Ca^{+2}, Mg^{+2} ions) such as ethylene diamine tetra acetic acid or its alkali metal salts. Buffers used to stabilize the latex toward pH changes such as phosphate or citrate salts [5, 20]. Chain transfer agents like mercaptans.

3. Surfactant free emulsion polymerization

Used for manufacture of adhesives and water resistant polymers. By absence of surfactant, intensive coagulation of the particles greatly reduces the number of particles per unit volume of water so, particle nucleation and growth reduced [1]. Several literature had been reported about surfactant free emulsion polymerization, in this section a brief hint about these publications will be considered. Tauer et al. [45] studied the surfactant-free emulsion polymerization of styrene initiated by KPS. Wang and Pan [46] studied the surfactant-free emulsion polymerization of styrene with the water soluble co-monomer as 4-vinylpyridine. Ni et al. [47] studied mechanism of particle nucleation through adding 8% ethyl acetate at low speed agitation (100–200 rpm) through polymerization of 4-vinyl pyridine and styrene. Ou et al. [48] investigated the effect of the hydrophilic co monomer (vinyl acetate or methyl methacrylate) on particle nucleation in the surfactant-free emulsion polymerization of styrene. Yan et al. [49] investigated the surfactant-free emulsion copolymerization of styrene, methyl methacrylate and acrylic acid initiated by ammonium persulfate. Other literature reported by Mahdavian and Abdollahi, Zhang et al., Shaffei et al., and Sahoo and Mohapatra [50–53].

3.1. Emulsion polymerization mechanism

Emulsion polymerization is a free radical polymerization protocol occurs in three distinct steps; initiation, propagation, and termination.

3.1.1. Initiation

In which the initiator decomposed to free radicals either by (1) hemolytic fission (hemolysis) through thermal decomposition or radiation and by (2) chemical reaction through redox reactions. Rate of initiator dissociation (R_d) is the rate determining step and given by Eqs. (1)–(3);

$$I \xrightarrow{K_d} 2R\bullet \tag{1}$$

$$R_d = 2fK_d[I] \tag{2}$$

$$R\bullet + M \xrightarrow{K_i} RM\bullet \tag{3}$$

Rate of initiation (R_i) is given by Eq. (4);

$$R_i = 2fK_i[I] \tag{4}$$

K_d rate constant for initiator dissociation

f Initiator efficiency

[I] Initiator concentration

K_i rate constant for initiation

3.1.2. Propagation

Involve continuous addition of monomer particles to active centers ($RM^\bullet$) to form polymer chains.

Rate of polymerization (R_p) given by Eq. (5);

$$R_p = -\frac{d[M]}{dt} = k_i[R\bullet][M] + k_p[M\bullet][M] \tag{5}$$

Where $[R^\bullet]$ is the free radicals concentration, [M] is the monomer concentration and $[M^\bullet]$ is the total concentration of active monomers. Since consumed monomers in initiation stage is very small as compared to propagation, so the term "$k_i[R\bullet][M]$" can be neglected and rate of polymerization is determined by rate of propagation; Eq. (6).

$$R_p = k_p[M\bullet][M] \tag{6}$$

http://dx.doi.org/10.5772/intechopen.72143

3.1.3. Termination

Termination leads to loss of two growing polymer chains [3]. It occurs by either recombination or disproportionation. Recombination involves reaction of one polymer chain with another growing one and reactive sites are blocked according to the following equation.

$$P_n \bullet + P_m \bullet \xrightarrow{K_t} P_{n+m} \tag{7}$$

Disproportionation where one chain abstract a hydrogen proton from another leaving it with unsaturated end group according to the following equation. This termination mechanism result in two polymer chain fractions one is saturated and the other is unsaturated [31].

$$P_n \bullet + P_m \bullet \xrightarrow{K_t} P_n + P_m \tag{8}$$

Termination may occur by chain transfer reactions, which involves removal of atom and formation of new radical which may initiate the reaction forming other segments or cannot initiate the reaction so, the propagation progress ceased [31]. Other literature reported about termination occur by addition of retarders or inhibitors like phenols and catechol's to terminate active sites [31, 54].

3.2. Kinetics of emulsion polymerization

Since rate of polymerization expressed by Eq. (9);

$$R_p = k_p[M\bullet][M] \tag{9}$$

Where $[M\bullet]$ expressed by Eq. 10;

$$[M\bullet] = \frac{N' n}{NA} \tag{10}$$

N concentration of micelles plus particles

n the average number of radicals per micelle plus particle

N_A is the Avogadro number

$$R_p = \frac{N' nkp[M]}{NA} \tag{11}$$

The value of "n" determine rate of polymerization and depend on radical diffusion out of the polymer particles (desorption), the particle size, modes of termination, and the rates of initiation and termination relative to each other and to the other reaction parameters [5]. Depending on "n" value there are three cases that can be summarized as;

3.2.1. Case 1: n = 0.5

Means that at any given moment half of the polymer particles contain one radical and are growing while the other half are dormant, and known as zero–one systems to indicate that a polymer particle contains either zero or one radical at any given moment [5].

3.2.2. Case 2: n < 0.5

In which radical desorption from particles and termination in the aqueous phase are low especially for small particle sizes and low initiation rates.

3.2.3. Case 3: n > 0.5

In which particle size is large or the termination rate constant is low while termination in the aqueous phase and the initiation rate is fast, as some polymer particles contain two or more radicals.

- Degree of polymerization (X_n) is defined as the rate of growth of a polymer chain divided by the rate at which primary radicals enter the polymer particle and given by the following Eq. (12);

$$X_n = \frac{r_p}{r_i} = \frac{NKp[M]}{R_i} \tag{12}$$

This equation neglect any termination by chain transfer, if chain transfer occur the degree of polymerization given by Eq. (13).

$$X_n = \frac{r_p}{r_i \sum r_t} \tag{13}$$

where, $\sum r_t$ is the sum of termination reactions by chain transfer.

- Number of polymer particles is dependent on the total surface area of surfactant present in the system and given by Eq. (14);

$$N = K\left(\frac{Ri}{\mu}\right)^{2/5}(a_s S)^{3/5} \tag{14}$$

a_s is the interfacial surface area occupied by a surfactant molecule

S is the total concentration of surfactant in the system (micelles, solution, monomer droplets)

μ is the rate of volume increase of polymer particle

The number of polymer particles can be increased by increasing the emulsifier concentration while maintaining a constant rate of radical generation.

4. Conclusion

Emulsion polymerization has wide application on academic and industrial applications. This chapter discuss importance of heterogeneous emulsion polymerization and its constituents; moreover, a comprehensive analysis of the kinetics of emulsion polymerization has been presented. Other emulsion polymerization characteristics like inverse emulsion, mini emulsion and micro emulsion discussed briefly.

Author details

Abdelaziz Nasr Moawed Bakr El-hoshoudy

Address all correspondence to: azizchemist@yahoo.com

Production Department, Egyptian Petroleum Research Institute, Production Department, Cairo, Egypt

References

[1] Chern CS. Emulsion polymerization mechanisms and kinetics. Progress in Polymer Science. 2006;**31**(5):443-486

[2] Silvério Neto W, Thyago Jensen A, Ribeiro Ferreira G, Fonseca Valadares L, Gambetta R, Belém Gonçalves S, et al. A survey on synthesis processes of structured materials for biomedical applications: Iron-based magnetic nanoparticles, polymeric materials and polymerization processes. Current Pharmaceutical Design. 2015;**21**(37):5336-5358

[3] Thickett SC, Gilbert RG. Emulsion polymerization: State of the art in kinetics and mechanisms. Polymer. 2007;**48**(24):6965-6991

[4] Rao JP, Geckeler KE. Polymer nanoparticles: Preparation techniques and size-control parameters. Progress in Polymer Science. 2011;**36**(7):887-913

[5] Odian G. Emulsion polymerization,Principles of polymerization. *Fourth edition* John Wiley & Sons, Inc; 2004:350-371

[6] Schild HG. Poly (N-isopropylacrylamide): Experiment, theory and application. Progress in Polymer Science. 1992;**17**(2):163-249

[7] Lee H, Poehlein G. Continuous tube-CSTR reactor system for emulsion polymerization kinetic studies. Chemical Engineering Science. 1986;**41**(4):1023-1030

[8] Landfester K. Polyreactions in miniemulsions. Macromolecular Rapid Communications. 2001;**22**(12):896-936

[9] López-Quintela MA, Tojo C, Blanco M, Rio LG, Leis J. Microemulsion dynamics and reactions in microemulsions. Current Opinion in Colloid & Interface Science. 2004;**9**(3):264-278

[10] Lopez-Quintela MA. Synthesis of nanomaterials in microemulsions: Formation mechanisms and growth control. Current Opinion in Colloid & Interface Science. 2003;**8**(2):137-144

[11] Abdou L, El-Molla M, Hakeim O, El-Gammal M, Shamey R. Synthesis of nanoscale binders through mini emulsion polymerization for textile pigment applications. Industrial & Engineering Chemistry Research. 2013;**52**(6):2195-2200

[12] Capek I, Chern C-S. Radical polymerization in direct mini-emulsion systems. New Polymerization Techniques and Synthetic Methodologies. Springer. 2001:101-165

[13] Carey V, Wang Y-G. Mixed-effects models in S and S-PLUS. Journal of the American Statistical Association, Taylor & Francis. 2001;(96):1135-1136

[14] Chern C-S. Emulsion polymerizations in nonuniform latex particles.Principles and applications of emulsion polymerization. John Wiley & Sons; 2008:200-222

[15] Liu G, Li L, Yang X, Dai Z. Preparation of silica/polymer hybrid microspheres and the corresponding hollow polymer microspheres with functional groups. Polymers for Advanced Technologies. 2008;**19**(12):1922-1930

[16] Van Berkel KY, Hawker CJ. Tailored composite polymer–metal nanoparticles by miniemulsion polymerization and thiol-ene functionalization. Journal of Polymer Science Part A: Polymer Chemistry. 2010;**48**(7):1594-1606

[17] Harkins WD. A general theory of the mechanism of emulsion polymerization 1. Journal of the American Chemical Society. 1947;**69**(6):1428-1444

[18] Smith WV, Ewart RH. Kinetics of emulsion polymerization. The Journal of Chemical Physics. 1948;**16**(6):592-599

[19] Gao J, Penlidis A. Mathematical modeling and computer simulator/database for emulsion polymerizations. Progress in Polymer Science. 2002;**27**(3):403-535

[20] Nicholson J. The chemistry of polymers. Royal Society of Chemistry. 2017

[21] Tardy A, Nicolas J, Gigmes D, Lefay C, Guillaneuf Y. Radical ring-opening polymerization: Scope, limitations, and application to (bio) degradable materials. Chemical Reviews. 2017;**117**(3):1319-1406

[22] Wang W, Meng L, Hackett R, Keefe D. Developmental ability of human oocytes with or without birefringent spindles imaged by Polscope before insemination. Human Reproduction. 2001;**16**(7):1464-1468

[23] Lander ES, Linton LM, Birren B, Nusbaum C, Zody MC, Baldwin J, et al. Initial sequencing and analysis of the human genome. 2001;(409):860-921

[24] Pan ZW, Dai ZR, Wang ZL. Nanobelts of semiconducting oxides. Science. 2001;**291**(5510):1947-1949

[25] El-hoshoudy A, Desouky S, Betiha M, Alsabagh A. Use of 1-vinyl imidazole based surfmers for preparation of polyacrylamide–SiO 2 nanocomposite through aza-Michael addition copolymerization reaction for rock wettability alteration. Fuel. 2016;**170**:161-175

[26] El-hoshoudy A, Desouky S, Alsabagh A, Betiha M, MY E-k, Mahmoud S. Evaluation of solution and rheological properties for hydrophobically associated polyacrylamide copolymer as a promised enhanced oil recovery candidate. Egyptian Journal of Petroleum. 2016;779-85

[27] El-Hoshoudy A, Desouky S, Elkady M, Alsabagh A, Betiha M, Mahmoud S. Investigation of optimum polymerization conditions for synthesis of cross-linked polyacrylamide-amphoteric surfmer nanocomposites for polymer flooding in sandstone reservoirs. International Journal of Polymer Science. 2015:2015

[28] El-hoshoudy A, Desouky S, Elkady M, Alsabagh A, Betiha M, Mahmoud S. Hydrophobically associated polymers for wettability alteration and enhanced oil recovery: Article review. Egyptian Journal of Petroleum:2016

[29] El-hoshoudy AN, Desouky SM, Betiha MH, Alsabagh AM. Hydrophobic Polymers Flooding. Application and Characterization of Surfactants. InTechopen; 2017:75-95

[30] El Hoshoudy A, Desouky S, Al-sabagh A, El-kady M, Betiha M, Mahmoud S. Synthesis and Characterization of Polyacrylamide Crosslinked Copolymer for Enhanced Oil Recovery and Rock Wettability Alteration 2015;**3**(4):47-59

[31] Yamak HB. Emulsion polymerization: effects of polymerization variables on the properties of vinyl acetate based emulsion polymers. Polymer Science. InTechopen; 2013:35-72

[32] Li X, Zhu D, Wang X, Wang N, Gao J, Li H. Thermal conductivity enhancement dependent pH and chemical surfactant for cu-H 2 O nanofluids. Thermochimica Acta. 2008;**469**(1):98-103

[33] Reb P, Margarit-Puri K, Klapper M, Müllen K. Polymerizable and nonpolymerizable isophthalic acid derivatives as surfactants in emulsion polymerization. Macromolecules. 2000;**33**(21):7718-7723

[34] Cochet F, Claverie JP, Graillat C, Sauterey F, Guyot A. Emulsion polymerization of chloroprene in the presence of a maleic polymerizable surfactant: Control of gel formation at low conversion. Macromolecules. 2004;**37**(11):4081-4086

[35] Summers M, Eastoe J. Applications of polymerizable surfactants. Advances in Colloid and Interface Science. 2003;**100**:137-152

[36] Wang Y, Wu F. Amphiphilic acrylamide-based copolymer with porphyrin pendants for the highly selective detection of hg 2+ in aqueous solutions. Polymer. 2015;**56**:223-228

[37] Xue W, Hamley IW, Castelletto V, Olmsted PD. Synthesis and characterization of hydrophobically modified polyacrylamides and some observations on rheological properties. European Polymer Journal. 2004;**40**(1):47-56

[38] Freedman HH, Mason JP, Medalia A. Polysoaps. II. The preparation of vinyl soaps1. The Journal of Organic Chemistry. 1958;**23**(1):76-82

[39] Summers M, Eastoe J, Richardson RM. Concentrated polymerized cationic surfactant phases. Langmuir. 2003;**19**(16):6357-6362

[40] Benbayer C, Saidi-Besbes S, de Givenchy ET, Amigoni S, Guittard F, Derdour A. Copolymerization of novel reactive fluorinated acrylic monomers with styrene: Reactivity ratio determination. Colloid and Polymer Science. 2014;**292**(7):1711-1717

[41] Samakande A, Hartmann PC, Cloete V, Sanderson RD. Use of acrylic based surfmers for the preparation of exfoliated polystyrene–clay nanocomposites. Polymer. 2007;**48**(6):1490-1499

[42] X-J X, Chen F. Semi-continuous emulsion copolymerization of butyl methacrylate with polymerizable anionic surfactants. Polymer. 2004;**45**(14):4801-4810

[43] Arz C. The use of nonionic polymerizable surfactants in latexes and paints. Macromolecular Symposia. 187. Wiley Online Library; 2002:199-206.

[44] Eliseeva VI, Ivanchev S, Kuchanov S, Lebedev A. Emulsion polymerization and its applications in industry. Springer Science & Business Media; 2012:3-24

[45] Tauer K, Deckwer R, Kühn I, Schellenberg C. A comprehensive experimental study of surfactant-free emulsion polymerization of styrene. Colloid & Polymer Science. 1999;**277**(7):607-626

[46] Wang Y-M, Pan C. Study of the mechanism of the emulsifier-free emulsion polymerization of the styrene/4-vinylpyridine system. Colloid & Polymer Science. 1999;**277**(7):658-665

[47] Ni H, Ma G, Nagai M, Omi S. Effects of ethyl acetate on the soap-free emulsion polymerization of 4-vinylpyridine and styrene. II. Aspects of the mechanism. Journal of Applied Polymer Science. 2001;**82**(11):2692-2708

[48] Ou J-L, Yang J-K, Chen H. Styrene/potassium persulfate/water systems: Effects of hydrophilic comonomers and solvent additives on the nucleation mechanism and the particle size. European Polymer Journal. 2001;**37**(4):789-799

[49] Ce Y, Cheng S, Feng L. Kinetics and mechanism of emulsifier-free emulsion copolymerization: Styrene-methyl methacrylate-acrylic acid system. Journal of Polymer Science Part A: Polymer Chemistry. 1999;**37**(14):2649-2656

[50] Mahdavian A-R, Abdollahi M. Investigation into the effect of carboxylic acid monomer on particle nucleation and growth in emulsifier-free emulsion copolymerization of styrene–butadiene–acrylic acid. Polymer. 2004;**45**(10):3233-3239

[51] Zhang J, Zou Q, Li X, Cheng S. Soap-free cationic emulsion copolymerization of styrene and butyl acrylate with comonomer in the presence of alcohols. Journal of Applied Polymer Science. 2003;**89**(10):2791-2797

[52] Shaffei K, Ayoub M, Ismail M, Badran A. Kinetics and polymerization characteristics for some polyvinyl acetate emulsions prepared by different redox pair initiation systems. European Polymer Journal. 1998;**34**(3):553-556

[53] Sahoo PK, Mohapatra R. Synthesis and kinetic studies of PMMA nanoparticles by non-conventionally initiated emulsion polymerization. European Polymer Journal. 2003;**39**(9):1839-1846

[54] Antonietti M, Tauer K. 90 years of polymer latexes and heterophase polymerization: More vital than ever. Macromolecular Chemistry and Physics. 2003;**204**(2):207-219

Chapter 2

Ring Opening Metathesis Polymerization

Alexey Lyapkov, Stanislav Kiselev,
Galina Bozhenkova, Olga Kukurina,
Mekhman Yusubov and Francis Verpoort

Additional information is available at the end of the chapter

http://dx.doi.org/10.5772/intechopen.71085

Abstract

In recent years, the olefins metathesis has established itself as a powerful tool for carbon-carbon bonds forming and has found numerous applications in polymer chemistry. One of the important directions of metathesis is the polymerization with cycle opening. A study of new ruthenium catalysts, resistant to the many functional groups effects, has showed the possibility of synthesizing functionalized polymers with unique properties. In this chapter, reactivity and activation parameters of eight different norbornene dicarboxylic acid alkyl esters in the presence of a Hoveyda-Grubbs II catalyst for the ring opening metathesis polymerization were determined by ^{1}H NMR analysis *in-situ*. The molecules of esters differ in the aliphatic radical structure and the location of the substituent groups. Kinetic studies have shown that effective polymerization constants and activation parameters strongly depend on the monomer structure. It is shown that the elongation of the aliphatic radical does not significantly affect the reactivity, but significantly changes the activation parameters. The branching of the aliphatic radical significantly affects both the reactivity of the corresponding ester and the activation parameters of the polymerization. The position of the substituents in the norbornene ring of the ester also has a significant effect on the activation parameters of metathesis polymerization.

Keywords: ring opening metathesis polymerization, nuclear magnetic resonance, dicyclopentadiene, alkyl esters of norbornene dicarboxylic acid, Hoveyda-Grubbs catalysts, observed rate constant, activation energy

1. Introduction

Ring opening metathesis polymerization (ROMP) is a process of one or more cyclic olefins transformation to polymer catalyzed by metal carbene compounds. Indeed, the number of double bonds both in polymer and in monomer is equal [1–4]. As well as in any other type of polymerization, ROMP can be divided into several stages (**Scheme 1**).

The initiation begins with the coordination of the cycloolefins' double bond with the metal-carbene complex. The next step is a formation of a metal-cyclobutane intermediate. This process occurs in each act of addition of the monomer during the chain growth. The intermediate decomposes to form a new metal-carbene complex, with the growing chain being a ligand attached to the metal via a double bond.

The polymerization continues until the monomer is completely reacted or an equilibrium state is reached or the reaction is terminated by the addition of a special reagent that blocks the catalyst.

Living polymerization with ring opening metathesis is terminated by removing the transition metal from the end of the growing chain and its further deactivation. Deactivation in this case involves the formation of a complex unable to initiate polymerization [5]. Ethyl vinyl ether is an effective stopper for most ruthenium catalysts. It forms a very stable complex of the [Ru]=CHOEt type and ensures the functionalization of the polymer end-group. Acrylate derivatives of 2-butene-1,4-diol, succinic anhydride or butyl acrylate can also stop the growth of the macromolecule [6].

ROMP, as well as most metathesis reactions, is reversible, so the described transformations can proceed both in the forward and backward directions. The direction of the reaction can be predicted using the Gibbs energy:

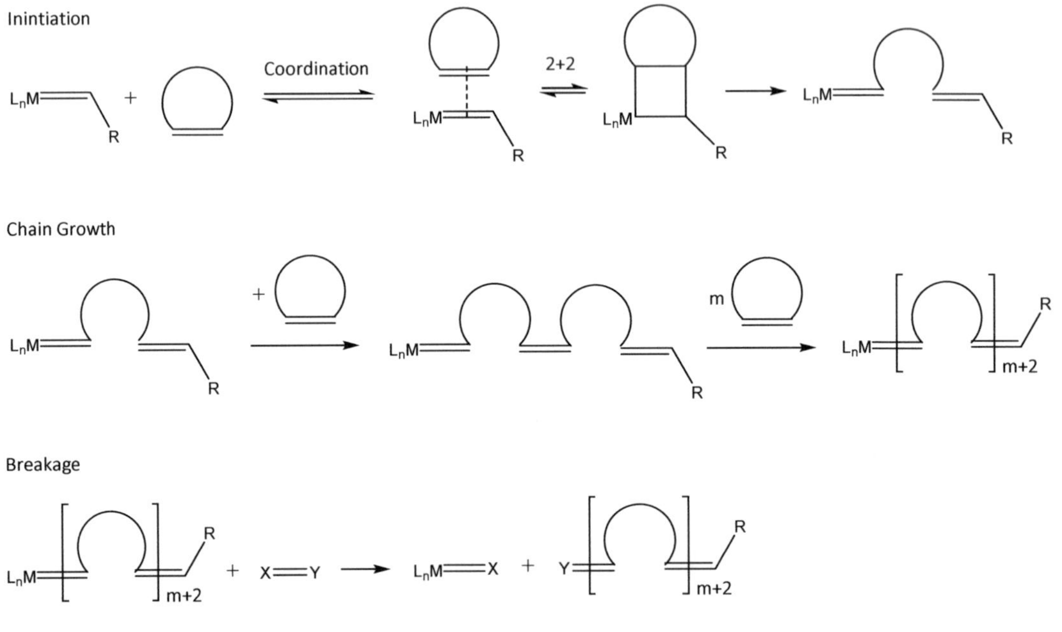

Scheme 1. Stages of ROMP.

$$\Delta G = -RT \cdot \ln K_{\text{равн}} = \Delta H - T \cdot \Delta S \tag{1}$$

The enthalpy factor makes an essential contribution to ROMP's Gibbs energy. This fact is explained by a high strain energy of cyclic unsaturated compounds participating in the polymerization [7]. The cycle strain energy of most cycloalkenes is more than 20 kJ/mole. For example, norbornene has an energy of about 100 kJ/mole [8]. The strain energy of the cycle releasing during the decomposition of the metal-cyclobutane complex and maintaining the forward direction of the reaction. However, if the deformation energy of cycle is low, the contribution of the entropy factor into the Gibbs energy becomes more significant in comparison with a smaller enthalpy factor. In this case, the entropy factor must be properly reduced in order to implement the direct process by increasing the monomer concentration or decreasing the temperature of the process [9].

Secondary metathesis reactions can occur in addition to the main described reactions during the polymerization process. Intermolecular and intramolecular chain transfers are two main side reactions of ROMP (**Scheme 2**).

During the intermolecular transferring, the active metal-carbene complex located at the end of one macromolecule interacts with the double bond of the adjoined macromolecule, which leads to fragment exchange process. The reaction proceeds simultaneously in two directions. One of them provides two polymer chains with the active ruthenium on both. The second leads to one inactive chain and one chain with two active centers.

At the intramolecular chain transferring, the active metal-carbene complex reacts with the double bond of the same macromolecule led to a cyclization of the polymer chain. The listed side reactions, eventually, lead to a broadening of the molecular mass distribution (MMD) and a decrease in a molecular weight of the polymer [10].

The chain transfer as well as the spontaneous termination of the growing chain is highly improbable, so ROMP is a living polymerization. ROMP polymers are characterized by high molecular weights and a narrow MMD, as well as for products of other living polymerizations [11].

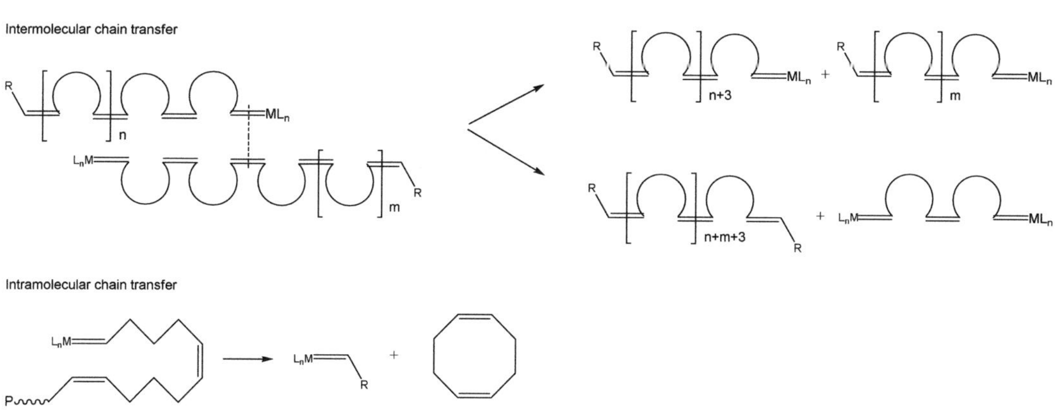

Scheme 2. Intermolecular and intramolecular chain transfers in ROMP.

2. Well-defined ruthenium metathesis catalysts

Up until the mid-1970s, the information about the structure of the active center of olefin metathesis catalysts was not known. Catalytic systems were mixtures of various compounds containing a transition metal. For example, in the late 1960s, Calderon and Goodyear employees published a number of papers about usage of a catalyst consisting of WCl_6, $AlEtCl_2$, and ethanol [12, 13]. Cyclic olefins form polymers (copolymers) of a high tacticity in a presence of this catalyst [14]. Catalytic systems prepared from compounds based on transition metals such as vanadium $VCl_4/Al(Hex)_3$; $V(Ac)_3/AlC1(C_2H_5)_2$; titanium $TiCl_4/Al(C_2H_5)_3$ and $TiCl_4/AlCl_3/Al(C_2H_5)_3$ [15]; molybdenum $MoCl_5/Al(C_2H_5)_3$ [16] as well as catalysts based on osmium, ruthenium, and iridium chlorides [17] were used in ROMP.

In 1976, through the example of the Fisher catalyst $WCPhR(CO)_5$ (where R=Ph or —OCH_3), described by Casey and Fisher [18, 19], Katz was the first who showed the ability of a metal-carbene complex to catalyze the process of metathesis polymerization independently without additional compounds [20]. In the history of metathesis, catalytic complexes with a well-studied structure were called "well-defined" catalysts. The discovery of "well-defined" catalysts had significantly increased the ability of ROMP to obtain polymers that have unique properties.

The first "well-defined" ruthenium catalyst was synthesized by Grubbs in 1992. The alkylidene source was 3,3-diphenylcyclopropene [21] (**Scheme 3**).

Unfortunately, this catalyst had a low activity in comparison with already available metathesis catalysts. Replacing triphenylphosphine ligands with tricyclohexylphosphine significantly improved the activity of the catalyst (**Figure 1**—1 and 2).

Later, in 1995, catalytic complexes known as first-generation Grubbs catalysts (**Figure 1**—3 and 4) were prepared using phenyl diazomethane. These catalysts not only had equal activity to molybdenum catalysts but also were indifferent to the polar groups in the monomer [22, 23].

In 1999, Grubbs reported the synthesis of second-generation catalysts (**Figure 1**—5), showing better activity and more stability at air. This catalyst was obtained by replacing tricyclohexylphosphine with an N-heterocyclic carbene ligand [24]. A year later, Hoveyda's group reported on a new type of catalytic system based on the catalysts of Grubbs of the first and second generations (**Figure 1**—6 and 7). These complexes include a chelating ester ligand [25]. Recently, a new type of ruthenium catalyst has appeared where the N-heterocyclic carbene

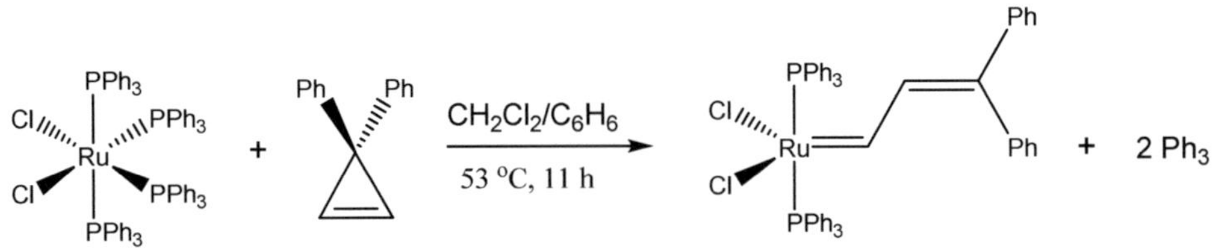

Scheme 3. Scheme for the Grubbs I catalyst synthesis.

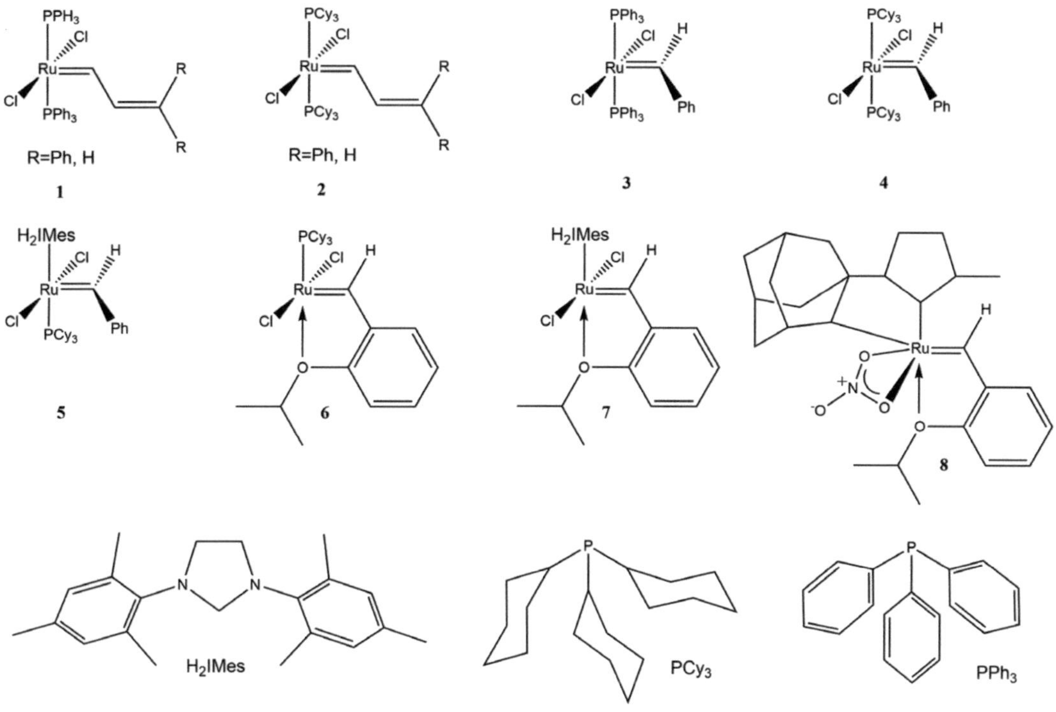

Figure 1. Main types of ruthenium catalysts.

ligand chelates the metal through the Ru-carbon bond (**Figure 1**—8). Such complexes possess high cis-selectivity in ROMP [26].

3. Reactivity of esters of 5-norbornene-2,3-dicarboxylic acid in ROMP

Currently, despite the fact that ethers of 2,3-norbornene dicarboxylic acid appear to be potential material for synthesizing polymers via ROMP, the interrelation between the molecule structure and their reactivity for metathesis polymerization with full ring opening has not been stated. A few polymerization mechanisms with different catalysts (including ruthenium) are known; however, there is no detailed description of how ethers of 5-norbornene-2,3-dicarboxylic acid behave. The study of reaction activity of 5-norbornene-2,3-dicarboxylic acid ethers with different structure using an appropriate catalyst (carbene complex of ruthenium (1,3-bis-(2,4,6-trimethylphenyl)-2-imidoazolidevynilidene)dichloro(ortho-N,N-dimethylaminomethyl-phenylmethylene)-ruthenium—**1** (**Figure 2**) [27] has filled this gap.

In this research, we used alkyl diesters of bicyclo[2.2.1]hept-5-en-2,3 dicarboxylic acid, obtained according to the technique given in the paper (**Figure 3**) [28].

Polymerization was carried out in NMR tubes, concurrently measuring the proton spectrum after a certain period using AU-program *zgser*.

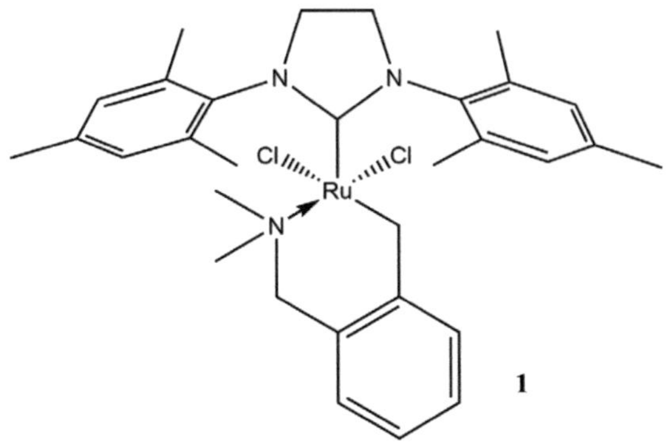

Figure 2. Catalyst complex of ruthenium used as an initiating agent for polymerization.

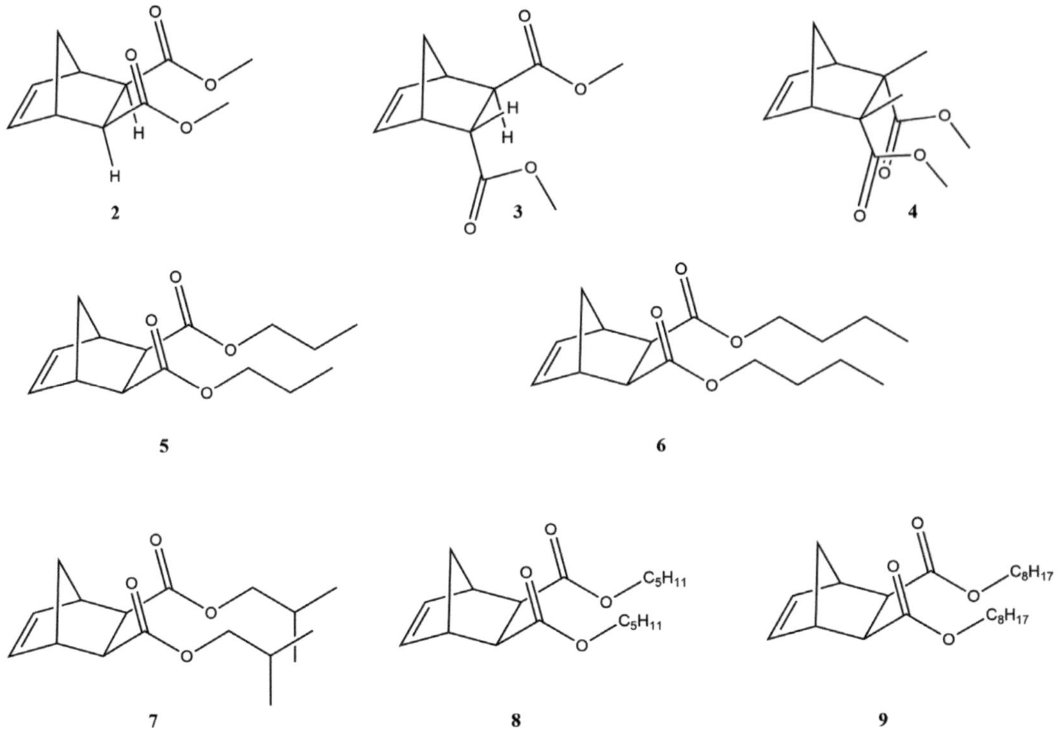

Figure 3. Alkyl diesters of bicyclo [2.2.1]hept-5-en-2,3 dicarboxylic acid, used as monomers.

The monomer concentrations were determined based on decrease and growth of integrated intensities of resonances of olefinic protons of monomer—S_M and polymer—S_P (**Figure 4**)

$$C_M = C_{M_0} \cdot \frac{S_M}{S_M + S_P} \tag{2}$$

$$C_K = C_{K_0} \cdot \frac{S_{K_0} \cdot S_S}{S_{S_0} \cdot S_K} \tag{3}$$

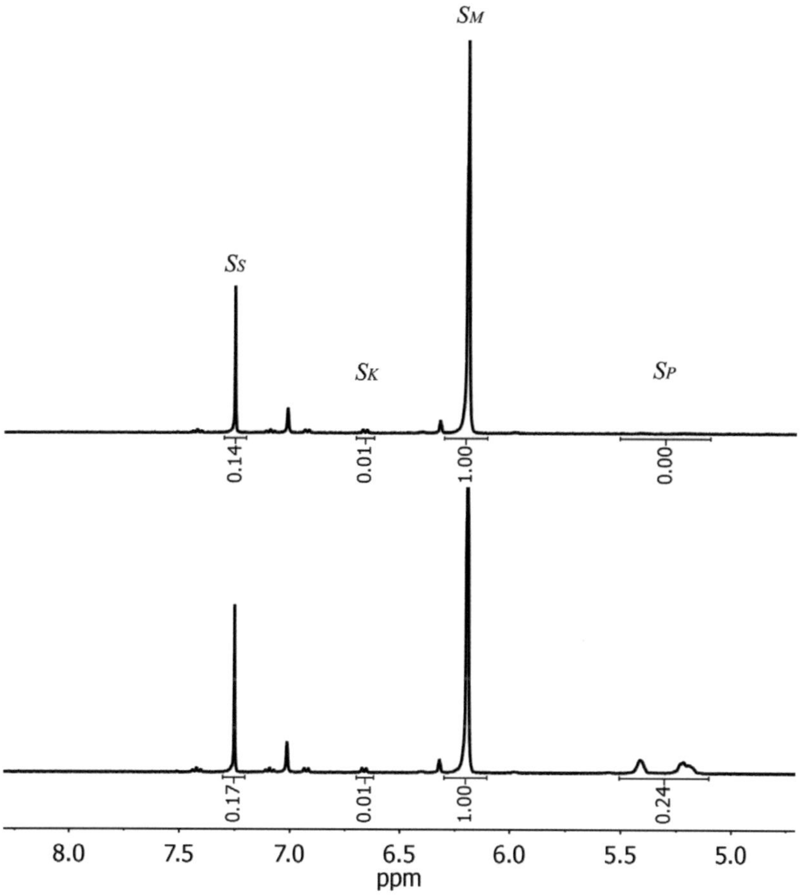

Figure 4. Combining the fragments of NMR spectra **2** in the beginning of a reaction and after 20 min.

where C_{K_0} and S_{S_0}—squares of integrated intensities of a catalyst and solvent, measured in the beginning of the reaction; S_K and S_S—current squares of integrated intensities of a catalyst and solvent during the reaction.

The AU-program *multintegr* was used to gauge the integrated intensities and time of the experiment. The using of low-viscosity solvents allowed obtaining high-resolution proton NMR spectra. Thus, kinetic studies should be carried out in the solution. And the set of monomer concentrations was defined to get kinetic correlations based on the spectral data. The solvent should be used as a diluent. **Figure 5(a)** demonstrates the curves describing the changes of concentration **2** in the course of time. According to the literature data, chloroform-d was taken as a solvent.

The molecules of chloroform-d do not react with active ruthenium and play a role of a polar medium, which stabilize 14-electron state of the active ruthenium [29]. Initially, toluene-d8 was suggested as a possible solvent, but the catalyst and monomers dissolve better in chloroform-d, which is also a widely used and more available solvent for NMR studies than toluene-d8.

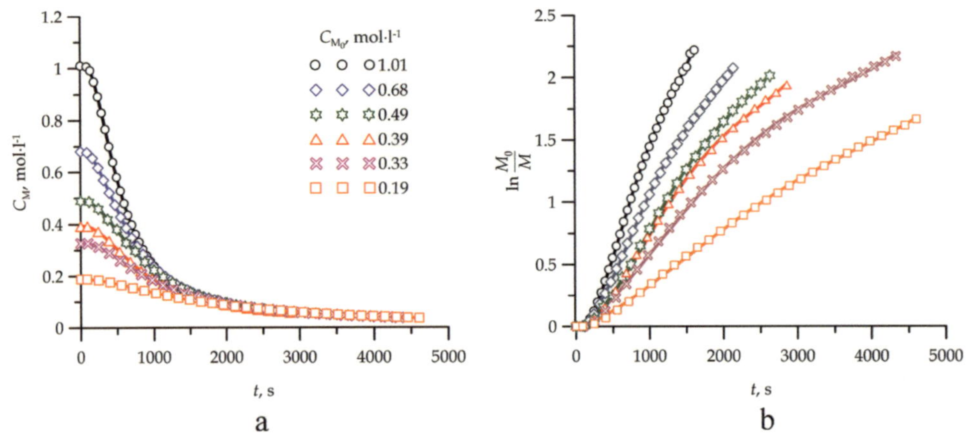

Figure 5. The decrease of monomer **2** (a) and its semi-logarithmic anamorphoses (b) during polymerization with catalyst 1 with varying initial concentration of the monomer (C_{K_0} = 0.0087 mole l^{-1}, 50°C).

Also, it was shown that reactivity of dimethyl ether of exo,exo-norbornene dicarboxylic acid is higher in chloroform-d [30]. Since chloroform-d boils at 60.9°C in ambient conditions, the operational temperature range was limited to 50°C to prevent any changes in the reactant concentration which could be caused by evaporation.

Studies [31, 32] considered the ring opening metathesis polymerization as pseudo first-order reaction as regards to the monomer concentration, which is valid for polymerization of above-mentioned monomers.

Figure 5(b) demonstrates that there can be seen three regions in the semi-logarithmic anamorphoses. The first region has non-linear segment of curve corresponding to the initiation stage. The second one is the straight-line segment prolongs to the extent of 70% monomer conversion (till one on the logarithmic scale, **Figure 5(b)**). The third region is a noticeable non-linear segment of curves, which is observed after 70% conversion. The appearance of such non-linear segments is due to the viscosity of the reaction mixture increasing, owing to the polymer molecular weight growth. This results to the fact that the polymerization rate is limited by the diffusion of monomer molecules to the active ruthenium.

Figure 6 shows the straight-line ranges of semi-logarithmic anamorphoses of polymerization **2**, catalyzed by **1**. The slope of the right lines corresponds to the observed constant of polymerization k_o.

Based on correlation coefficients given in **Table 1**, we can conclude that semi-logarithmic anamorphoses are linear in the noticed interval. **Figure 6** shows the correlation of the constant k_o and initial monomer concentration.

Figure 7 shows that k_o linearly depends on the monomer concentration within the following range from 0.2 to 1.0 mole l^{-1}, which allows to vary the monomer concentration in this range to implement kinetic experiments.

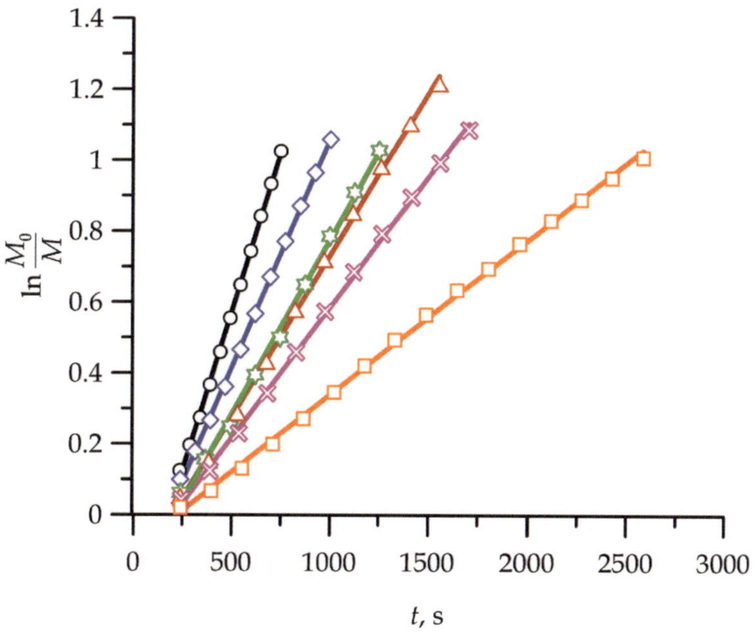

Figure 6. Linear segments of semi-logarithmic anamorphoses of monomer **2** polymerization over catalyst **1** (C_{K_0} = 0.0087 mole l^{-1}, 50°C, dependences are marked in accordance with **Figure 5**).

C_{M_0}, mole l^{-1}	a	b	r	$10^3 \cdot k_o$
1.01	0.00183	−0.3417	0.999	1.83
0.69	0.00131	−0.2416	0.999	1.31
0.49	0.00094	−0.1992	0.999	0.94
0.39	0.00089	−0.1614	0.999	0.89
0.33	0.00070	−0.1116	0.998	0.70
0.19	0.00042	−0.0751	0.999	0.42

Table 1. Values of k_o which calculated out of linear dependences on **Figure 6**.

Ruthenium complex should be activated to initiate polymerization. This is carried out by the first addition of monomer, which is initiation stage as well. There exist several possible mechanisms of activation; however, based on the literature data, it is assumed that bulky olefins, including the research monomers, interact with active ruthenium on a dissociative mechanism [33]

$$K \underset{k_{-1}}{\overset{k_1}{\rightleftarrows}} K^* \xrightarrow{+M\ k_2} P^*$$

The initiation rate equals the rate of active centers formation P^*. The active centers formation occurs in two stages. As it can be seen from **Figure 8**, the concentration of ruthenium complex slightly changes.

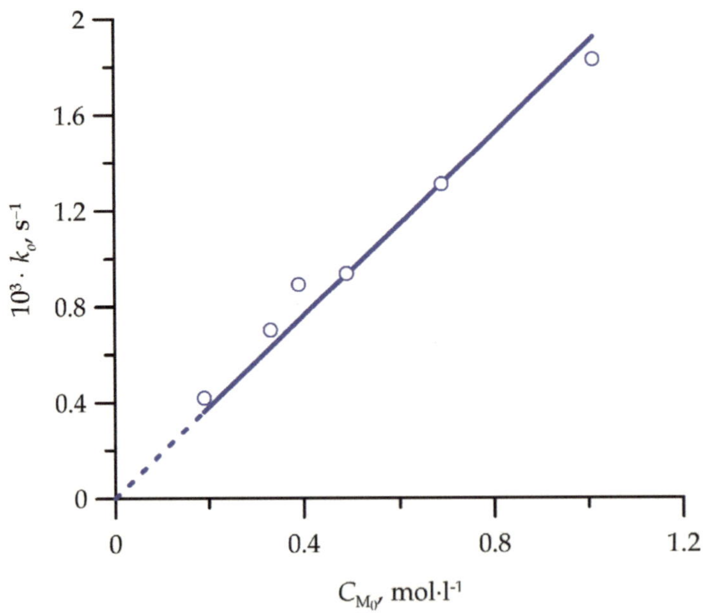

Figure 7. Correlation of the observed constant k_o of monomer **2** polymerization, catalyzed by **1** with the initial monomer concentration (C_{K_0} = 0.0087 mole l^{-1}, 50°C).

Its decrease is 1–2% from the initial catalyst concentration. Synthesized polymers at these conditions possess high molecular weight (**Table 2**).

Based on **Figure 8** and **Table 2**, we can conclude that the formation of active centers is slower than the growth of polymer chain. The research [34] also confirmed this, stating that for polymerization of exo-exo-5,6-bi(methoxycarbonyl)-7-oxabicyclo[2.2.1]hept-2-ene over Grubbs catalyst of the first generation the correlation of constants is k_i/k_g = 0.23. Moreover, the study [29] suggests that the correlation k_i/k_g is even lesser and equals 0.03 for catalyst with N-chelating ligand. In addition, based on the data presented, we can assume that disassociating of nitrogen defined by constant k_1 is limiting in the initiation reaction. Notably that the monomer molecule does not interact during initiation, that is why the formation rate of the active ruthenium complex K^* only depends on the temperature and initial concentration of ruthenium complex. Thus, the structure of the monomer molecule can affect the second stage of initiation defined by constant k_2 and the stage of polymer chain growth defined by constant k_g (it is suggested that constants of different stages of polymerization are equal $k_g^1 = k_g^2 = \ldots = k_g$)

$$P^* + M \xrightarrow{k_g^1} P^*M$$
$$P^*M + M \xrightarrow{k_g^2} P^*M_2$$
$$\ldots$$
$$P^*M_n + M \xrightarrow{k_g} P^*M_{n+1}$$

http://dx.doi.org/10.5772/intechopen.71085

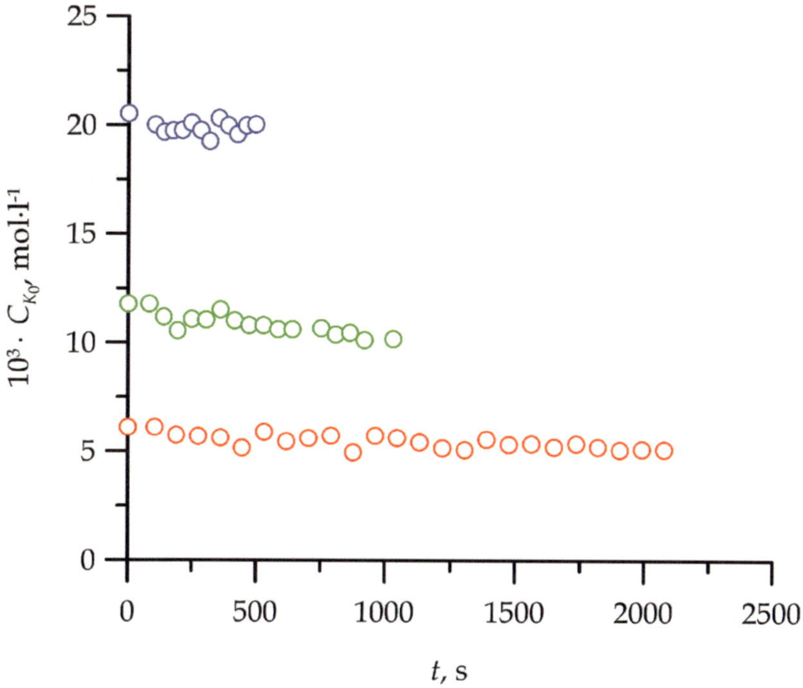

Figure 8. Ruthenium complex decrease during polymerization of monomer **2** with different initial concentration of catalyst **1** (C_{M_0} = 0.35 mole l^{-1}, 50°C).

C_{K_0}, mole l^{-1}	C_{M_0}/C_{K_0}	$10^{-5} \cdot M_n$, g mole^{-1}
0.021	17	8.5
0.012	30	10.1
0.006	57	7.6

Table 2. Average molecular weight of the obtained polymers depending on the number of initial reagents.

Kinetics of monomer consumption is complicated (**Figures 5(a)** and **9**).

In polymerization, a monomer is used during initiation and growth of the polymer chain

$$-\frac{dC_M}{dt} = k_2 \cdot C_{K^*} \cdot C_M + k_g \cdot C_{P^*} \cdot C_M \qquad (4)$$

The concentration of active ruthenium complex C_{K^*} and concentration of active chains C_{P^*} are low, with C_{P^*} due to the absence of reactions of termination [5, 6] and transfer [22] of the chain constantly increases during the reaction. Since k_1 is much lesser than constants k_{-1} and k_2, it is possible to apply the principle of quasistationary for concentration of the active form C_{K^*}:

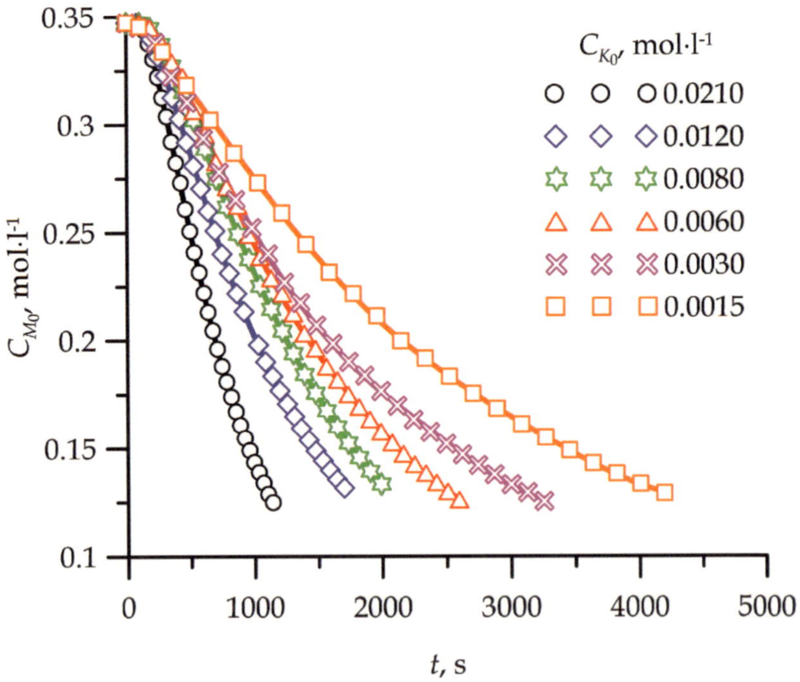

Figure 9. Monomer **2** consumption in the polymerization reaction over varying initial catalyst concentration C_{M_0} = 0.35 mole l^{-1}, 50°C).

$$\frac{dC_{K^*}}{dt} = k_1 \cdot C_K - k_{-1} \cdot C_{K^*} - k_2 \cdot C_{K^*} \cdot C_M = 0 \quad (5)$$

$$C_{K^*} = \frac{k_1 \cdot C_K}{k_{-1} - k_2 \cdot C_M} \quad (6)$$

The second stage of the initiation reaction can be viewed as pseudo first-order one proceeding with effective constant $k_{2_e} = k_2 \cdot C_{M_0}$. This assumption is fair as $C_{K_0} \ll C_{M_0}$ and $C_{K_0} \gg C_{K^*}$. Taking into consideration that the catalyst concentration slightly changes during the reaction, it could be considered that $C_K \cong C_{K_0}$. Then, changes in the concentration of active chains over time are defined by the following equation:

$$\frac{dC_{P^*}}{dt} = \frac{k_1 \cdot C_{K_0}}{k_{-1} - k_2 \cdot C_{M_0}} \cdot k_2 \cdot C_{M_0} \quad (7)$$

After integrating we get:

$$C_{P^*} = \frac{\frac{k_1}{k_{-1}} \cdot k_2 \cdot C_{K_0} \cdot C_{M_0}}{1 + \frac{k_2}{k_{-1}} \cdot C_{M_0}} \cdot t \quad (8)$$

The amount of active chains is equal to the decreasing of the monomer, which is forming these chains. Knowing the active chains' concentration from Eq. (8), the change in monomer concentration in time can be described by Eq. (9):

$$-\frac{dC_M}{dt} = \frac{\frac{k_1}{k_{-1}} \cdot k_2 \cdot C_{K_0} \cdot C_{M_0}}{1 + \frac{k_2}{k_{-1}} \cdot C_{M_0}} + \frac{\frac{k_1}{k_{-1}} \cdot k_2 \cdot C_{K_0} \cdot C_{M_0}}{1 + \frac{k_2}{k_{-1}} \cdot C_{M_0}} k_g \cdot C_M \cdot t \qquad (9)$$

To simplify the equation and implement semi-logarithmic coordinates for defining the rate constant, we can ignore the first component of the right side of the equation, since it contributes less if compared with the second component. This assumption is fair for the later stages of polymerization. The formula $\frac{\frac{k_1}{k_{-1}} \cdot k_2 \cdot C_{K_0} \cdot C_{M_0}}{1+\frac{k_2}{k_{-1}} \cdot C_{M_0}} \cdot t$ can be expressed as the following product $C_{K_0} \cdot f$, where f is the effectiveness of initiation equal to $\frac{C_{P*}}{C_{K_0}} = \frac{\frac{k_1}{k_{-1}} \cdot k_2 \cdot C_{M_0} \cdot t}{1+\frac{k_2}{k_{-1}} \cdot C_{M_0}}$. Then, we can put it down the following way:

$$-\frac{dC_M}{C_M} = k_g \cdot C_{K_0} \cdot f \cdot dt \qquad (10)$$

After integrating, we would acquire:

$$\ln \frac{C_{M_0}}{C_M} = k_g \cdot C_{K_0} \cdot f \cdot t \qquad (11)$$

Taking into consideration that f for each monomer differs only by the value of k_2 constant, which depends on the structure of monomer, it is possible to compare reaction capacity and values of activation parameters using product $f \cdot k_g$.

The chain growth rate constant of polymer k_g times the effectiveness of initiation f corresponds the tangent of the slope in the straight-line segment of semi-logarithmic correlation, which equals the product of the observed constant k_o times the initial catalyst concentration C_{K_0} (**Figure 10**).

Correlations in **Figures 7** and **11** demonstrate that the observed constant of polymerization k_o linearly depends on both the initial concentration of monomer and the initial catalyst concentration.

Linear correlation of k_o from C_{K_0} is observed because Eq. (1) takes the initial concentration of catalyst into consideration. In turn, k_o linearly depends on C_{M_0} since Eq. (11) includes parameter f, which depends on the initial concentration of monomer. Based on the data presented, we can conclude that it is possible to use the effective constant $k_e = \frac{k_o}{C_{K_0} \cdot C_{M_0}}$ to compare reaction capacity of the ethers under study. The dimensionality of constant k_e correspond the dimensionality of second-order constant since the concentration of monomer is included in numerator

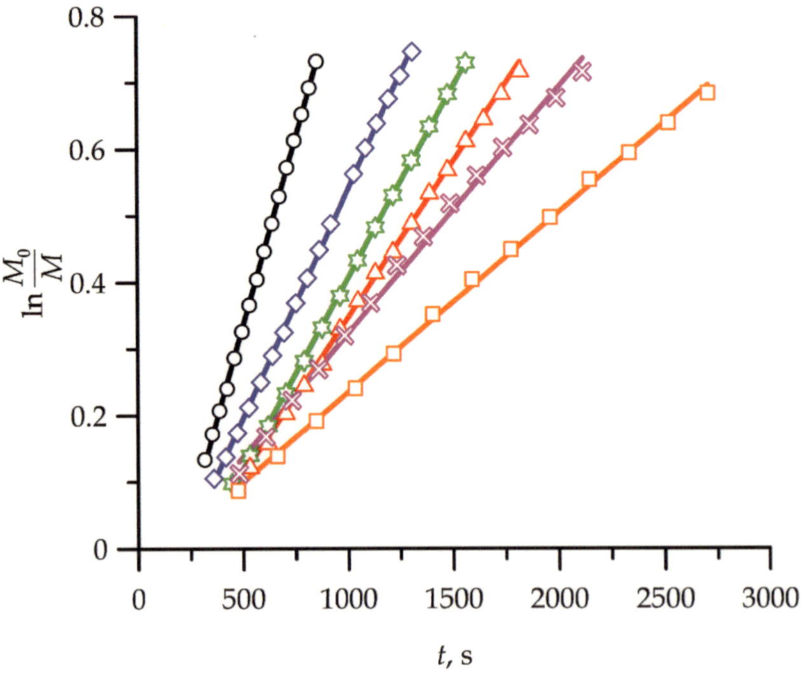

Figure 10. Semi-logarithmic correlations of polymerization **2** over catalyst **1** with varying catalyst concentration (C_{M_0} = 0.35 mole·l^{-1}, 50°C, dependences are marked in accordance with **Figure 9**).

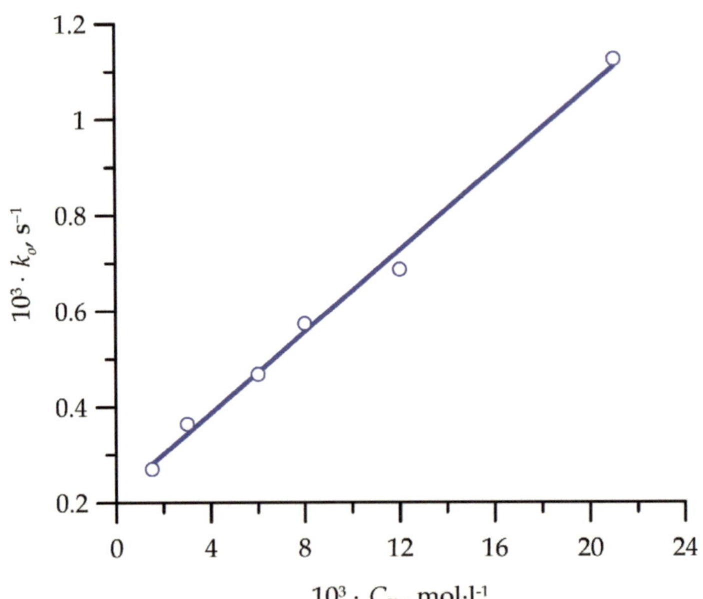

Figure 11. Correlation of the observed constant k_o of polymerization of monomer **2,** catalyzed by **1** with the initial concentration of catalyst (C_{M_0} = 0.35 mole l^{-1}, 50°C).

http://dx.doi.org/10.5772/intechopen.71085

and denominator of the equation of the initiation effectiveness. Since constant k_o depends on the initial concentration of catalyst linearly, we can use the noticed range of concentration to estimate reactivity of esters.

4. Reactivity-structure relationship of esters of 2,3-norbornene dicarboxylic acid

Based on the values of effective constants, we compared reactivity and activation parameters of polymerization of diesters exo,exo-2,3-norbornene dicarboxylic acid, which are differ by length and branched chain of ester substituent. To define the activation parameters, we used Arrhenius equation (12) and calculation results are shown in **Figure 12**

$$\ln k_e = \ln A - \frac{E_a}{R \cdot T} \tag{12}$$

This correlation between ln k_e and $1/T$ for each researched ester has linear character. This proves that the interaction mechanism of ruthenium complex and corresponding ester at the different temperatures is unchanged. **Table 3** presents data on effective constants and activation parameters of polymerization of diesters exo,exo-2,3-norbornene dicarboxylic acid.

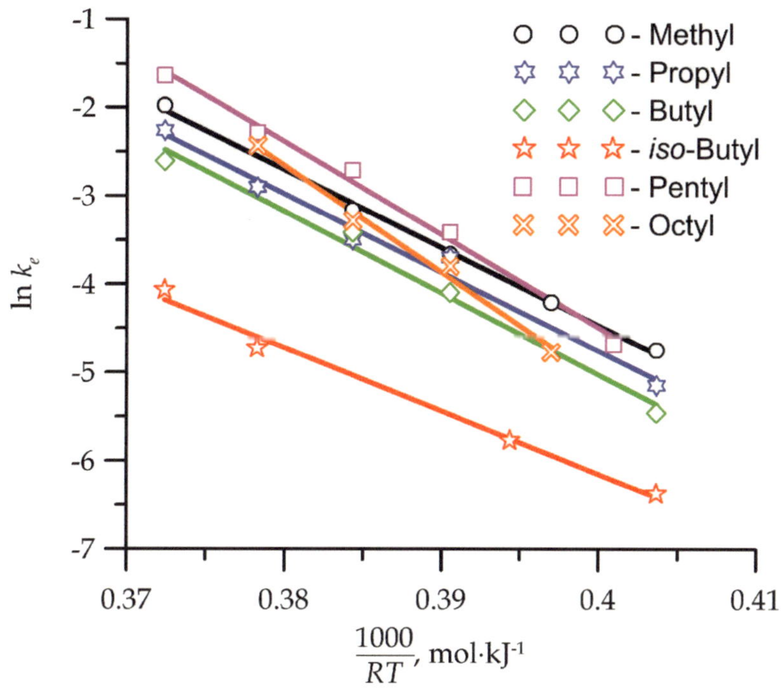

Figure 12. Arrhenius correlations of polymerization of diesters exo,exo-5-norbornene-2,3-dicarboxylic acid.

Substituent	k_e, l mole^{-1} s^{-1} (30°C)	E_a, kJ mole^{-1}	A, l mole^{-1} s^{-1}
Methyl	0.11	82	2×10^{12}
Propyl	0.10	89	2×10^{13}
Butyl	0.08	92	7×10^{13}
Iso-butyl	0.01	72	6×10^{9}
Pentyl	0.21	105	2×10^{16}
Octyl	0.17	121	2×10^{18}

Table 3. Effective constants and activation parameters of polymerization of diesters exo,exo-2,3-norbornene dicarboxylic acid.

It was expected that aliphatic radical elongation from the first to the eighth atoms of carbon would lead to gradual decrease in reactivity of esters row. However, according to **Table 5**, aliphatic radical elongation insignificantly affects the reactivity of esters.

On the contrary, branched substituent chain affects reactivity greatly. Constant k_e of an ester with iso-butyl radical is six times lesser than constant k_e of a similar ester with linear butyl radical. The steric hindrances significantly decrease the reactivity of diesters with branching aliphatic radical under interaction with active form of ruthenium complex. In study [35], the researchers attempted to make a quantitative estimation of the initiation and growth constants.

As shown in **Table 5**, the increase of aliphatic radical length leads to gradual increase of activation parameters. To explain changes in activation parameters, we should define the rate constant, which is dependent from monomer structure in more degree. Effective constant of polymerization includes four true constants. Constants k_1 and k_{-1} are determined by the structure of ruthenium complex and do not depend on the monomer structure. Nevertheless, the influence of the ester structure could be indirect. When bond Ru-N is disassociated, a 14-electron state is formed. This state is more polar than the initial 16-electron state (**Scheme 4**).

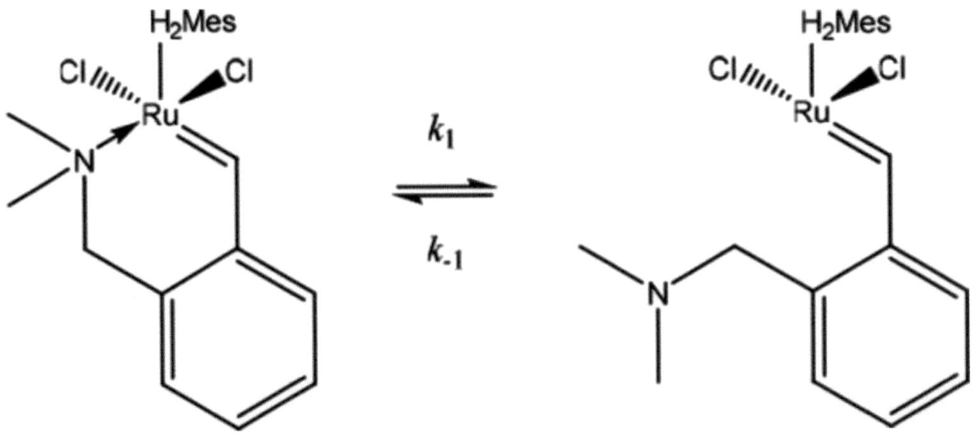

Scheme 4. Dissociation the Ru-N bond of catalyst.

Polar media stabilize 14-electron state and make disassociation easier. Solutions of esters, which are different in structure, may possess different dielectric permittivity and, thus, could affect constant k_1 and k_{-1}. However, in polymerization, solutions of esters have low concentration and the contribution of ester into the polarity of medium remains insignificant.

Esters structure would affect more constants k_2 and k_g. First, we should understand the way monomer structure can affect constant k_2. This constant defines the reaction rate, which identifies the process of monomer addition to the active form of ruthenium complex. In this process, the double bond of ester molecule occupies the vacant position in the coordination sphere of ruthenium complex (**Scheme 5**).

While the activation energy E_a defines excess of energy, which molecules in the reaction should possess to form transition state. Pre-exponential factor A can correlate with steric factor. Both parameters define the process of reaching the top of a potential barrier and are calculated from the initial state of the system. It is unlikely that the length of aliphatic radical affected the rate and activation parameters of this reaction. It is more probable that constant k_2 and activation parameters are nearly equal for molecules with varying length of aliphatic radical. It is also unlikely that branching substituent can affect both the rate and activation parameters of this process.

It is necessary to mention that the influence of the previous monomer unit may affect rate and activation parameters of monomer addition reaction to one of the active forms of ruthenium. However, this factor is absent on this stage of the reaction.

Having analyzed the experimental data, we concluded that the structure of monomer is more likely to affect the growth reaction of polymer chain with constant k_g.

It is known from literature data that esters of 5-norbornene-2,3-dicarboxylic acid can chelate the active forms of ruthenium complex with carbonyl oxygen of ester group, thus, forming hexatomic intramolecular complex [36]. Therefore, two active forms of ruthenium complex can take part in the polymer chain-growth reaction (**Figure 13**).

Ru—O bond strength depends on donor properties of carbonyl oxygen. In esters row, the donor properties of oxygen will enhance as there will increase inductive effect of growing radical. At the same time, Ru—O bond strength will increase. Reinforcement of Ru—O bond decreases mobility of ester fragment and makes its intramolecular complex more rigid.

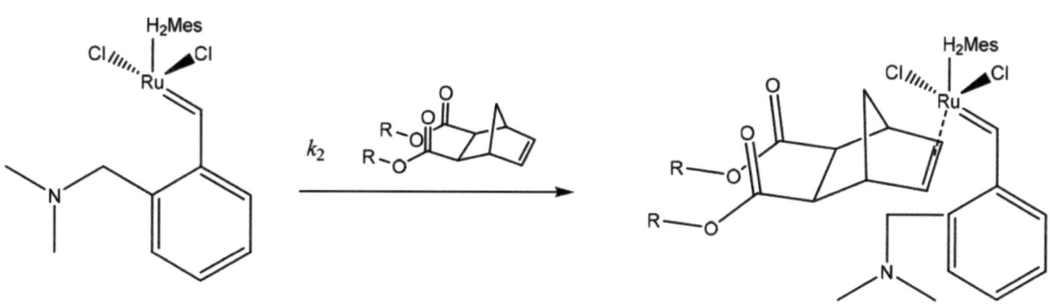

Scheme 5. Coordination of the monomer's molecule with ruthenium complex.

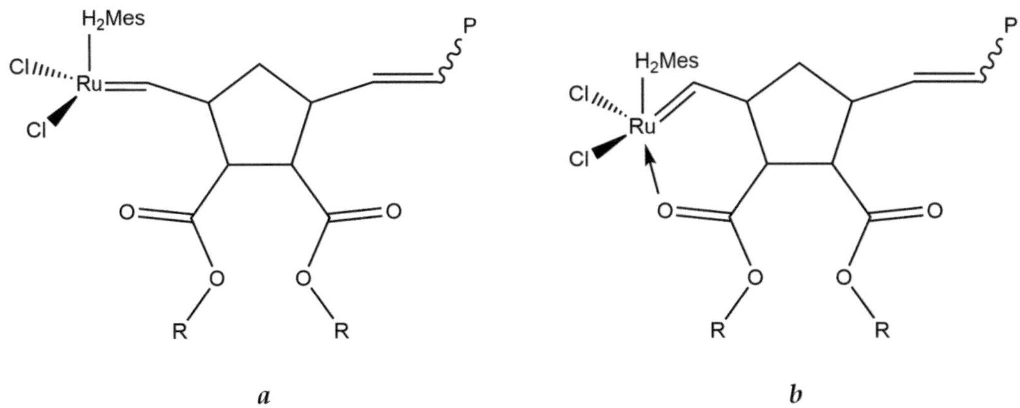

Figure 13. Non-chelated (a) and chelated (b) active forms of ruthenium complex.

When transition state is formed, monomer molecules occupy the position of oxygen in coordination sphere of ruthenium, what is accompanied by destruction of intramolecular complex (**Scheme 6**).

To degrade Ru—O bond, it is necessary to spend some energy. Lengthening of aliphatic radical, which promotes improvement of donor properties of carbonyl oxygen and intensification of Ru—O bond, increases the amount of energy needed to degrade Ru—O bond. That is why activation energy rises as the length of aliphatic radical increases. If the activation energy corresponds to the excessive energy that reacting molecules should possess to pass the potential barrier, then pre-exponential multiplier defines peculiarities of interaction of these molecules. Pre-exponential multiplier can correlate with the change of activation entropy, which depends on changes in the number of freedom degrees of the reacting molecules. Ruthenium and the previous monomer unit can form a ring with lesser number of freedom degrees than the complex they form of non-ring structure. Besides the rigidness of intramolecular complex depends on Ru—O bond strength (the more strength Ru—O bond, the more stable is intramolecular complex). Therefore, the increase of pre-exponential multiplier defined by the growth of aliphatic radical is explained by the increase in the number of freedom degrees, which appear when intramolecular complex degrades during the formation of transition state.

To form Ru—O bond, carbonyl oxygen and ruthenium should be positioned in a certain way. When Ru—O if formed, the molecule geometry is changed. Steric factor is one of the

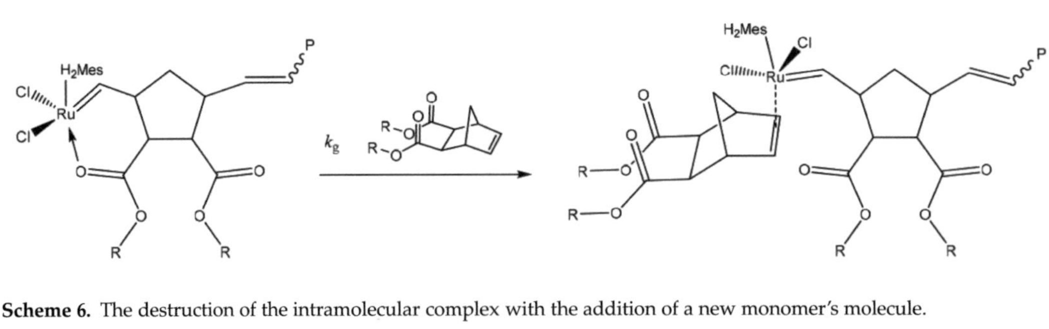

Scheme 6. The destruction of the intramolecular complex with the addition of a new monomer's molecule.

hindrances making the formation of intramolecular complex harder. In the case of ester with branched substituent, bulky iso-butyl radicals cannot set near each other properly for carbonyl oxygen to form strength bond with ruthenium due to steric hindrances. This reduces the activation energy and pre-exponential multiplier. In addition, iso-butyl fragments of the previous monomer unit hinder the monomer placement in the coordination sphere of ruthenium, which cuts reactivity of this ester.

Figure 14 demonstrates Arrhenius correlations of constant k_e in three 3-dimensional isomers of dimethyl ester of 5-norbornene-2,3-dicarboxyl acid. The correlations are linear in the range of temperatures, which proves that the interaction mechanism of ruthenium complex and the corresponding ester is permanent.

Based on the correlations in **Figure 14**, we calculated effective constants and activation parameters of polymerization. The results are in **Table 4**.

Table 4 shows that the orientation of ester substituents to the norbornene ring affects both reactivity and activation parameters of polymerization.

The presence of substituent in endo-position reduces reaction capacity of ester. This corresponds with the data shown in other studies [34, 37–40], which estimated reaction capacity of endo- and exo-isomers of dicyclopentadiene and 2,3-dicarbomethoxy-5-norbornene. In the research of Delaude at al. [40] measured the initiation constants for monomers **2, 3**, and **4** over $[RuCl_2(p\text{-}cymene)]_2$ complex activated with trimethylsilyldiazomethane; their values at 25°C were 0.040,

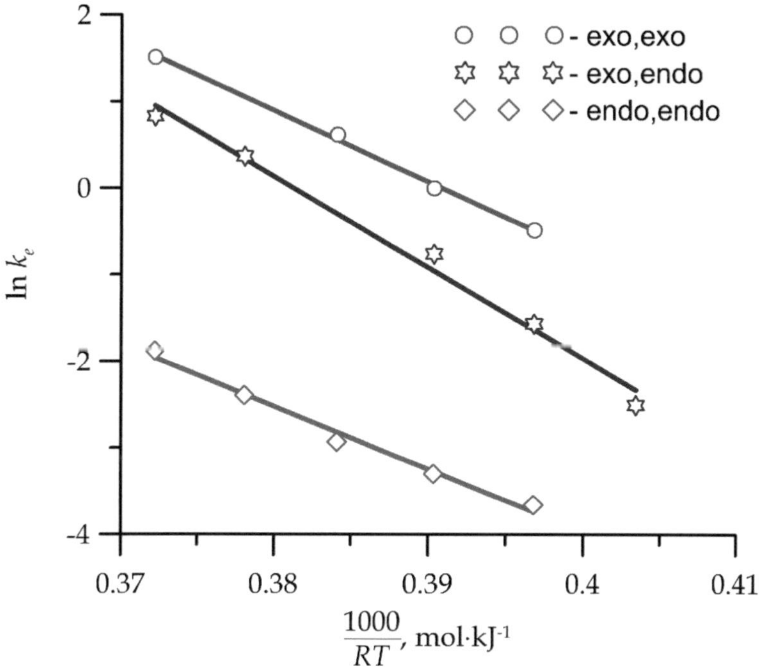

Figure 14. Arrhenius correlations of polymerization of three-dimensional isomers of dimethyl ester 5-norbornene-2,3-dicarboxylic acid.

Position	k_e, l mole^{-1} s^{-1} (30°C)	E_a, kJ mole^{-1}	A, l mole^{-1} s^{-1}
Exo,exo-	0.47	82	9×10^{13}
Exo,endo-	0.20	105	2×10^{17}
Endo,endo-	0.02	72	7×10^{10}

Table 4. Effective constants and activation parameters of polymerization of three-dimensional isomers of dimethyl ester 2,3-norbornene dicarboxylic acid.

0.025, and 0.05 l mole^{-1} s^{-1} for **2**, **3**, and **4**, respectively. At the same time, the constant of chain growth remains the same for all monomers and is within the range of 0.003–0.006 l mole^{-1} s^{-1}. The initiation stage of monomer **2** catalyzed by **1** is much slower than the chain growth stage. If we compare the constant of initiation and growth of monomer **2** catalyzed by **1** and [$RuCl_2$(p-cymene)]$_2$, then we would notice that initiation catalyzed by **1** is slower than chain growth in distinction from [$RuCl_2$(p-cymene)]$_2$, which affects initiation in a way that it is 10-fold faster than the growth of polymer chain. Comparison of constants defining polymerization initiated by these complexes is not adequate since these complexes have different structure and may have different activation mechanisms. However, in both the cases, ester groups in endo-position are located not far enough from the double bond of norbornene ring and sterically hinder the monomer attack by double bond of ruthenium. This would affect the polymerization rate of these esters both in the case of initiation by complex **1** and in the case of initiation by [$RuCl_2$(p-cymene)]$_2$. In the first case, the steric factor would affect both constants k_2 and k_g.

Each monomer is determined by its own set of activation parameters different from others. To explain the way the activation parameters change, we compiled a set of monomers in ascending order to form Ru—O bond and intramolecular complex. Exo,endo-isomer is more prone to form Ru—O bond since its ester substituents are located on different sides in relation to the norbornene ring and do not hinder each other during the formation of intramolecular complex. Ru—O bond is more strength, and intramolecular complex is more rigid in comparison with other isomers. That is why high activation energy and pre-exponential multiplier are typical for exo,endo-isomer. Exo,exo-isomer is the second on the capability to form Ru—O bond. This ester is inferior to exo,endo-isomer, since its ester substituents are located on one side in relation to norbornene ring. This sterically hinders their mutual distribution necessary for the formation of Ru—O bond. Ru—O bond has less strength, and intramolecular complex is more flexible. That is why if compared with exo,endo-isomer, exo,exo-isomer is defined by lower activation energy and pre-exponential multiplier.

Endo,endo-isomer is the third on the ability to form Ru—O bond. Because of the way ester substituents are located inside norbornene ring, this ester cannot form strong Ru—O bond. Ester group in endo-position cannot properly distribute in the coordination sphere of ruthenium to form intramolecular complex. That is why this molecule possesses low activation energy and pre-exponential multiplier.

In the paper [41], the authors estimated reactivity of these esters using the observed polymerization constant k_o as the criterion for comparing reaction capacity of monomers.

http://dx.doi.org/10.5772/intechopen.71085

5. Structure of polymers

The products of chemical reaction are no less valuable as a source of information for kinetic parameters. Their structure helps us to learn how reaction components interact with each other. Using NMR method to study the kinetics of metathesis polymerization of norbornene acid, ester is more beneficial since it allows estimating the structure of the obtained polymers immediately [25, 39, 42–44].

To analyze structure of polymers **2**, **5–8**, we used data from the study [39], which demonstrated that cis-units have resonances of olefinic protons in a stronger field in relation to trans-units (**Figure 15(a)**).

Overlap of resonances corresponding to cis- and trans-structures occurs in polymer obtained from monomer **4** (**Figure 15(c)**).

The spectrum of polymer **3** is more complex if compared with spectra of polymers **2** and **4**, since molecule **3** possesses chiral properties. To correlate the shifts, we applied the approach suggested in the following study [43]; it was used to analyze the structure of polymers obtained from chiral products of norbornene using NMR-spectra COSY.

Implementation of this approach alongside with assumption that resonances of olefinic protons in cis-fragments are shifted to a higher field in relation to trans-fragments [44] allowed referring resonances of olefin region to four possible structures (**Figures 15(b)** and **16**).

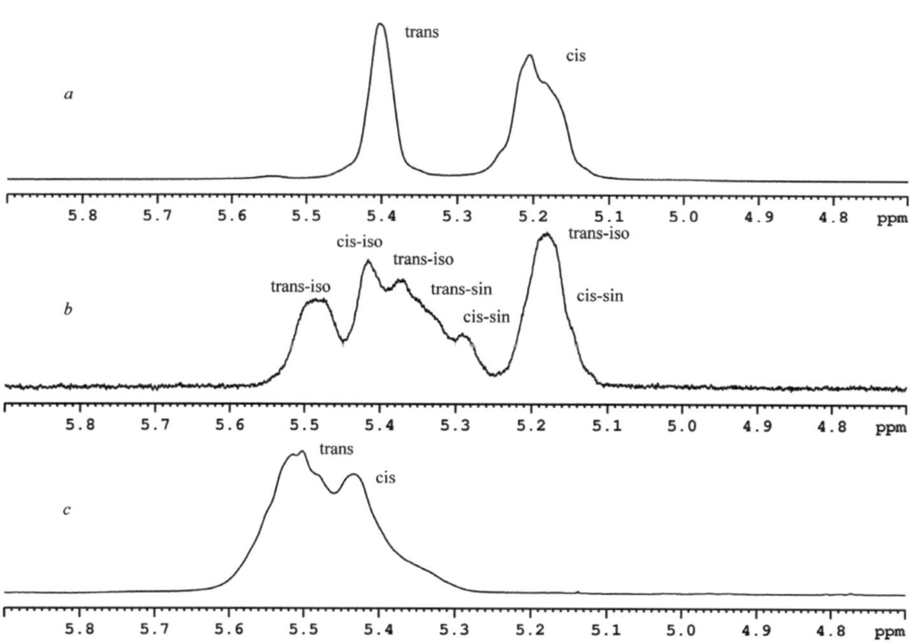

Figure 15. The region of olefinic protons ^1H NMR-spectra of polymers obtained with polymerization of **2–4** catalyzed by **1**.

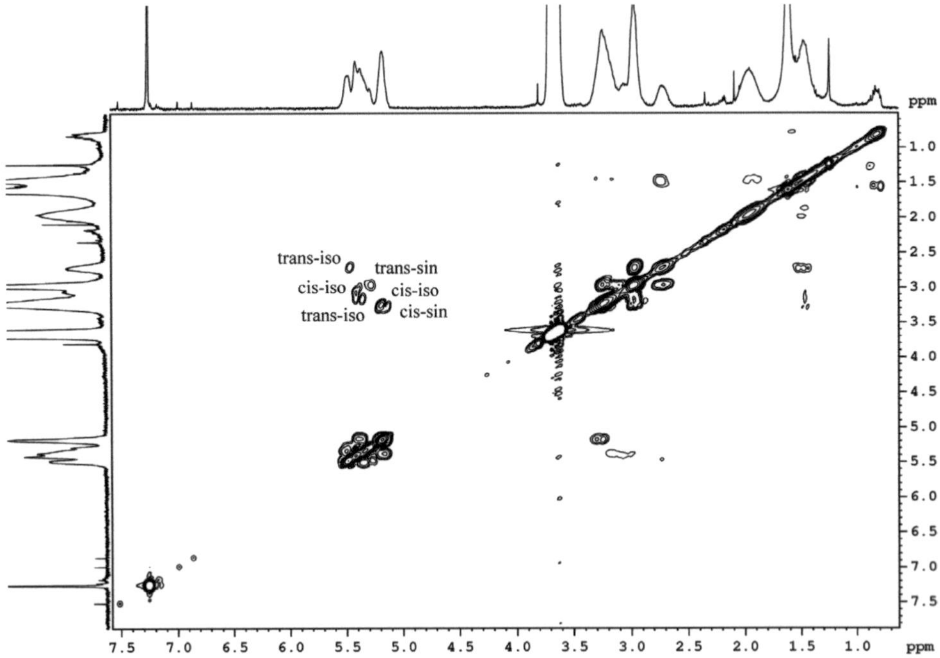

Figure 16. COSY-spectrum of polymer exo,endo-2,3-dicarbomethoxy-5-norbornene.

Monomer	**2**	**3**	**4**	**5**	**6**	**7**	**8**
Number of cis-units in polymer, %	57	54	43	56	55	56	55
Number of trans-units in polymer, %	43	44	57	44	45	44	45

Table 5. The number of cis- and trans-units in polymers obtained during polymerization of **2–8** over **1**.

Table 5 demonstrates what of cis- and trans-units of polymers obtained from diesters of 5-norbornene-2,3-dicarboxylic acid contain.

Data given in **Table 5** only offer estimative characteristic of polymers structure but allow comparing in series monomers under study. Given these data, we can highlight that polymers obtained from exo,exo-2,3-dicarbomethoxy-5-norbornenes have a similar structure. Neither elongation of radical of ester substituent nor its branching affects the ratio of cis- and trans-fragments. The change of substituents orientation in positions 2 and 3 in relation to norbornene ring causes the change in the number of cis- and trans-structures in the case of monomer **4**. Transfer of one ester substituent from exo- into endo-position would not bring about the increase of trans-units. The situation observed can be explained if we take into consideration that there are two ways monomer molecules are attached to active ruthenium with the formation of trans- and cis-structures (**Figure 17**).

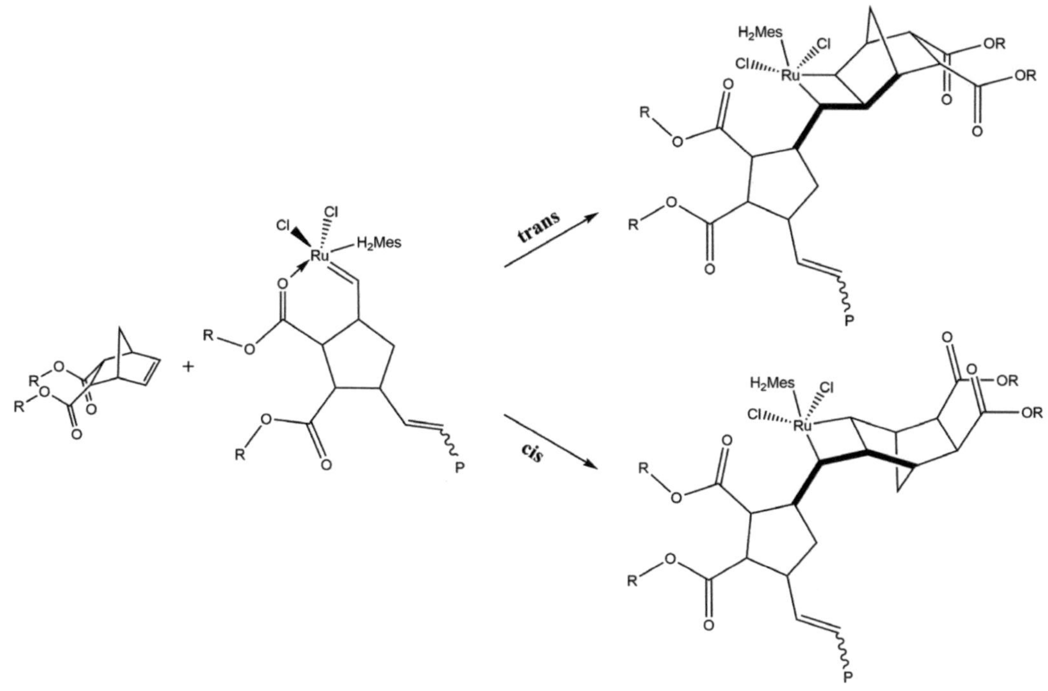

Figure 17. Two possible orientations **4** when attached to active form of ruthenium complex.

In the case when it is attached with the formation of trans-structure, a methylene bridge of norbornene ring and bulky H_2IMes-ligand hinder the monomer placement near the double bond of ruthenium. For exo,exo-derivatives, it is a more substantial hindrance if compared with ester groups which deter the attachment with the formation of cis-structures. On the contrary, for monomer **4**, two atoms of oxygen in esters are more of an obstacle for the attachment to active ruthenium than a methylene bridge of norbornene ring.

Figure 17 demonstrates that both carbonyl oxygen hinder the distribution of monomer **4** near the double bond of active ruthenium in such a way that attachment with the formation of trans-unit is sterically more beneficial. This is also seen in an increased number of trans-units in polymer obtained with monomer **4**. For monomer **3**, only one carbonyl oxygen is a hindrance and that is why the part of trans-units in the obtained polymer remains practically the same if compared with monomer **2**.

Thus, using monomers **2**, **5–8**, it is stated that the length and branching aliphatic radical of exo, exo-derivatives do not affect the ratio of cis- and trans-fragments in the obtained polymers. The orientation of ester substituents in relation to norbornene ring in 2,3-dicarbomethoxy-5-norbornenes affect the ratio of cis- and trans-fragments in polymers obtained from monomers **2–4**. The transfer of two ester substituents to endo-position increases the share of trans-units, which is due to more substantial steric hindrances caused by carbonyl oxygen of ester monomer group and H_2IMes-ligand of catalyst when forming cis-structure if compared with the obstacles caused by methylene bridge of norbornene ring and H_2IMes-ligand of catalyst when forming trans-structure.

6. Conclusion

To summarize, it is necessary to note that metathesis polymerization with cycle opening has proved to be a powerful method of synthesizing polymers. Materials obtained with this method possess good exploitation characteristics and have already proved effective on the market of polymer goods. Development of carbene complexes based on ruthenium has made it possible to synthesize polymers from ester of 5-norbornene-2,3-dicarboxylic acid using ROMP. To obtain polymer materials by described process, a technology of injection molding in which polymerization rate matters most. Largely, polymerization rate is defined by the structure of catalyst and monomer. The activity of ruthenium catalyst complexes is well studied and apart from the structure of the complex itself it depends on a number of external factors including temperature, solvent polarity, presence of acceptor or donor compounds, etc. The structure of monomer also affects polymerization rate. It is known that steric factor during polymerization of dicyclopentadiene and oxygenated derivatives of norbornene contribute greatly to the reaction rate.

In most studies, the researchers used kinetic correlations to estimate reaction capacity of different compounds. As a rule, kinetic data are obtained with NMR method, studying polymerization *in-situ*. The technique of such experiments is well adjusted by many scholars and has proved to be effective.

Implementing this approach, it is shown how the structure of esters of 5-norbornene-2,3-dicarboxylic acid affects their reaction capacity and activation parameters in metathesis polymerization with cycle opening initiated by ruthenium complex of Hoveyda-Grubbs type II with N-chelating ligand. Taking the values of activation parameters, it is assumed that there may exist active ruthenium in chelated form, which is proved in the following studies [28, 36, 45].

It was established that the increase in the length of hydrocarbon radical does not affect greatly the reactivity, but it influences substantially the activation parameters. Branching aliphatic radical affects greatly both reactivity and activation parameters of polymerization. Based on the change in activation parameters, it might be assumed that active form of ruthenium complex forms intramolecular complex with different stability of this complex.

It was stated that mutual position of ester substituents in relation to norbornene ring affects both reactivity and activation parameters of 2,3-dicrabomethoxy-5-norbornenes. The presence of a substituent for the monomer molecule in endo-position reduces reaction capacity of this monomer [46]. Activation parameters are directly depended on the ability of monomer to form intramolecular complex with active form of ruthenium.

Based on NMR-spectra, we estimated the structures of the obtained polymers. Based on the correlation of cis- and trans-fragments, it was established that the length and branching aliphatic radical of exo,exo-2,3-dicarboxy-5-norbornenes and exo,endo-orientation of ester substituents of 2,3-dicarbomethoxy-5-norbornene do not affect the structure of the obtained polymers. Endo,endo-orientation of two ester substituents of 2,3-dicarbomethoxy-5-norbornene increases the number of trans-units in the polymer, which is attributed to more substantial steric hindrances caused by carbonyl oxygen of monomer ester group and H_2IMes-ligand of catalyst when forming cis-structure if compared with hindrances caused by methylene bridge of norbornene ring and H_2IMes-ligand of catalyst when forming trans-structure.

Author details

Alexey Lyapkov[1]*, Stanislav Kiselev[2], Galina Bozhenkova[2], Olga Kukurina[1], Mekhman Yusubov[1] and Francis Verpoort[1,3]

*Address all correspondence to: alexdes@tpu.ru

1 National Research Tomsk Polytechnic University, Tomsk, Russia

2 Tomsk Oil and Gas Research and Design Institute, Tomsk, Russia

3 State Key Laboratory of Advanced Technology for Materials Synthesis and Processing; and Department of Applied Chemistry, Faculty of Sciences, Wuhan University of Technology, Wuhan, China

References

[1] Grubbs RH, Chang S. Recent advances in olefin metathesis and its application in organic synthesis. Tetrahedron. 1998;**54**(18):4413-4450. DOI: 10.1016/S0040-4020(97)10427-6

[2] Ivin KJ. Some recent applications of olefin metathesis in organic synthesis: A review. Journal of Molecular Catalysis A: Chemical. 1998;**133**(1):1-16. DOI: 10.1016/S1381-1169(97)00249-5

[3] Ahmed SR, Bullock SE, Steven E, Cresce SE, Kofinas P. Polydispersity control in ring opening metathesis polymerization of amphiphilic norbornene diblock copolymers. Polymer. 2003;**44**(17):4943-4948. DOI: 10.1016/S0032-3861(03)00487-7

[4] Bozhenkova GS, Ashirov RV, Lyapkov AA, Kiselev SA, Yusubov MS, Verpoort F. Novel polymers based on dimethyl esters of norbornene dicarboxylic acids synthesized using metathesis ring-opening polymerization. Current Organic Synthesis. 2017;**14**(3):383-388. DOI: 10.2174/1570179413666161031151319

[5] Bielawski CW, Grubbs RH. Living ring-opening metathesis polymerization. Progress in Polymer Science. 2007;**32**(1):1-29. DOI: 10.1016/j.progpolymsci.2006.08.006

[6] Lexer C, Saf R, Slugovc C. Acrylates as termination reagent for the preparation of semi-telechelic polymers made by ring opening metathesis polymerization. Journal of Polymer Science Part A: Polymer Chemistry. 2009;**47**(1):299-305. DOI: 10.1002/pola.23137

[7] Benson SW, Cruickshank FR, Golden DM, Haugen GR, O'Neal HE, Rodgers AS, Shaw R, Walsh R. Additivity rules for estimation of thermodynamical properties. Chemical Reviews. 1969;**69**(3):279-324. DOI: 10.1021/cr60259a002

[8] North M. ROMP of Norbornene Derivatives of Amino-Esters and Amino-Acids. In: Khosravi E., Szymanska-Buzar T. (eds) Ring Opening Metathesis Polymerisation and Related Chemistry. NATO Science Series (Series II: Mathematics, Physics and Chemistry), Springer, Dordrecht; 2002;**56**:157-166. DOI: 10.1007/978-94-010-0373-5_13

[9] Patton PA, Lillya CP, McCarthy TJ. Olefin metathesis of cyclohexene. Macromolecules. 1986;**19**(4):1266-1268. DOI: 10.1021/ma00158a056

[10] Benedicto AD, Claverie JP, Grubbs RH. Molecular weight distribution of living polymerization involving chain-transfer agents: Computational results, analytical solutions, and experimental investigations using ring-opening metathesis polymerization. Macromolecules. 1995;**28**(2):500-511. DOI: 10.1021/ma00106a013

[11] Darling TR, Davis TP, Fryd M, Gridnev AA, Haddleton DM, Ittel SD, Mathenson RR, Moad G, Rizzardo E. Living polymerization: Rationale for uniform terminology. Journal of Polymer Science Part A: Polymer Chemistry. 2000;**38**(10):1706-1708. DOI: 10.1002/(SICI)1099-0518(20000515)38:10<1706::AID-POLA20>3.0.CO;2-5

[12] Calderon N, Chen HY, Scott KW. Olefin metathesis—A novel reaction for skeletal transformations of unsaturated hydrocarbons. Tetrahedron Letters. 1967;**8**(34):3327-3329. DOI: 10.1016/S0040-4039(01)89881-6

[13] Calderon N, Ofstead EA, Ward JP, Judy WA, Scott KW. Olefin metathesis. I. Acyclic vinylenic hydrocarbons. Journal of the American Chemical Society. 1968;**90**(15):4133-4140. DOI: 10.1021/ja01017a039

[14] Calderon N, Ofstead EA, Judy WA. Ring-opening polymerization of unsaturated alicyclic compounds. Journal of Polymer Science Part A: Polymer Chemistry. 1967;**5**(9):2209-2217. DOI: 10.1002/pol.1967.150050901

[15] Dall'Asta G, Mazzanti G, Natta G, Porri L. Anionic coordinated polymerization of cyclobutene. Die Makromolekulare Chemie. 1962;**56**(1):224-227. DOI: 10.1002/macp.1962.020560118

[16] Natta G, Dall'Asta G, Mazzanti G. Stereospecific homopolymerization of cyclopentene. Angewandte Chemie, International Edition. 1964;**3**(11):723-729. DOI: 10.1002/anie.196407231

[17] Rinehart RE, Smith HP. The emulsion polymerization of the norbornene ring system catalyzed by noble metal compounds. Journal of Polymer Science, Part B: Polymer Letters. 1965;**3**(12):1049-1052. DOI: 10.1002/pol.1965.110031215

[18] Fischer EO, Maasböl A. On the existence of a tungsten carbonyl carbene complex. Angewandte Chemie, International Edition. 1964;**3**(8):580-581. DOI: 10.1002/anie.196405801

[19] Casey CP, Burkhardt TJ. (Diphenylcarbene)pentacarbonyltungsten(0). Journal of the American Chemical Society. 1973;**95**(17):5833-5834. DOI: 10.1021/ja00798a103

[20] Katz TJ, Lee SJ, Acton N. Stereospecific polymerization of cycloalkenes induced by a metal–carbene. Tetrahedron Letters. 1976;**17**(47):4247-4250. DOI: 10.1016/0040-4039(76)80086-X

[21] Nguyen ST, Johnson LK, Grubbs RH, Ziller JW. Ring-opening metathesis polymerization (ROMP) of norbornene by a group VIII carbene complex in protic media. Journal of the American Chemical Society. 1992;**114**(10):3974-3975. DOI: 10.1021/ja00036a053

[22] Schwab P, France MB, Ziller JW, Grubbs RH. A series of well-defined metathesis catalysts–synthesis of [$RuCl_2(=CHR')(PR_3)_2$] and its reactions. Angewandte Chemie, International Edition. 1995;**34**(18):2039-2041. DOI: 10.1002/anie.199520391

[23] Grubbs RH, Wenzel AG, editors. Handbook of Metathesis. Volume 1: Catalyst Development and Mechanism. Second ed. Weinheim: Wiley-VCH Verlag GmbH & Co. KGaA; 2015. p. 448

[24] Scholl M, Trnka TM, Morgan JP, Grubbs RH. Increased ring closing metathesis activity of ruthenium-based olefin metathesis catalysts coordinated with imidazolin-2-ylidene ligands. Tetrahedron Letters. 1999;**40**(12):2247-2250. DOI: 10.1016/S0040-4039(99)00217-8

[25] Garber SB, Kingsbury JS, Gray BL, Hoveyda AH. Efficient and recyclable monomeric and dendritic Ru-based metathesis catalysts. Journal of the American Chemical Society. 2000;**122**(34):8168-8179. DOI: 10.1021/ja001179g

[26] Keitz BK, Fedorov A, Grubbs RH. Cis-selective ring-opening metathesis polymerization with ruthenium catalysts. Journal of the American Chemical Society. 2012;**134**(4):2040-2043. DOI: 10.1021/ja211676y

[27] Afanasyev VV, Nizovtsev AV, Dolgina TM, Bespalova NB. Ruthenium catalyst for dicyclopentadiene polymerization and preparation method thereof (options). Pat. 2374269 RU, IPC C08F32/08, C08F132/08, C08F4/80, B01J27/13, B01J27/24. № 2008100385/04. PubDate: 27-11-2009; Applicant and patent holder of JSC SIBUR Holding

[28] Nelson WL, Freeman DS, Sankar R. Bicyclic amino alcohols. Isomeric 2-dimethylaminomethyl-3-hydroxymethylbicyclo[2.2.1]hept-5-enes. The Journal of Organic Chemistry. 1975;**40**(25):3658-3664. DOI: 10.1021/jo00913a010

[29] Sanford MS, Love JA, Grubbs RH. Mechanism and activity of ruthenium olefin metathesis catalysts. Journal of the American Chemical Society. 2001;**123**(27):6543-6554. DOI: 10.1021/ja010624k

[30] Kiselev SA, Zemlyakov DI, Ashirov RV, Lyapkov AA. Activity of original Hoveyda-Grubbs II catalyst in ring-opening metathesis polymerization of exo,exo-2,3-dicarbomethoxy-5-norbornene in two different solvents. In: Proceedings of III International Scientific School-Conference for Young Scientists "Catalysis: From Science to Industry"; 26–30 October 2014; Tomsk, Russia; Tomsk: Publishing "Ivan Fedorov"; 2014. p. 61

[31] Allaert B, Dieltiens N, Ledoux N, Vercaemst C, Van Der Voort P, Stevens CV, Linden A, Verpoort F. Synthesis and activity for ROMP of bidentate Schiff base substituted second generation Grubbs catalysts. Journal of Molecular Catalysis A: Chemical. 2006;**260**(1–2): 221-226. DOI: 10.1016/j.molcata.2006.07.006

[32] Holland MG, Griffith VE, France MB, Desjardins SG. Kinetics of the ring-opening metathesis polymerization of a 7-oxanorbornene derivative by Grubbs catalyst. Journal of Polymer Science Part A: Polymer Chemistry. 2003;**41**(13):2125-2131. DOI: 10.1002/pola.10761

[33] Louie J, Grubbs RH. Highly active metathesis catalysts generated *in situ* from inexpensive and air-stable precursors. Angewandte Chemie. 2001;**113**(1):253-255. DOI: 10.1002/1521-3757(20010105)113:1<253::AID-ANGE253>3.0.CO;2-Z

[34] Bazan GC, Schrock RR, Cho HN, Gibson VC. Polymerization of functionalized Norbornenes employing Mo(CH-t-Bu) (NAr) (O-t-Bu)$_2$ as the initiator. Macromolecules. 1991;**24**(16):4495-4502. DOI: 10.1021/ma00016a003

[35] Kiselev SA, Lenev DA, Lyapkov AA, Semakin SV, Bozhenkova GS, Verpoort F, Ashirov RV. Reactivity of norbornene esters in ring-opening metathesis polymerization initiated by a N-chelating Hoveyda II type catalyst. RSC Advances. 2016;**6**(7):5177-5183. DOI: 10.1039/C5RA25197D

[36] Haigh DA, Kenwright AM, Khosravi E. Nature of the propagating species in ring-opening metathesis polymerizations of oxygen-containing monomers using well-defined ruthenium initiators. Macromolecules. 2005;**38**(18):7571-7579. DOI: 10.1021/ma050838c

[37] P'Pool SJ, Schanz H-J. Reversible inhibition/activation of olefin metathesis: A kinetic investigation of ROMP and RCM reactions with Grubbs' catalyst. Journal of the American Chemical Society. 2007;**129**(46):14200-14212. DOI: 10.1021/ja071938w

[38] Rule JD, Moore JS. ROMP reactivity of endo- and exo-dicyclopentadiene. Macromolecules. 2002;**35**(21):7878-7882. DOI: 10.1021/ma0209489

[39] Ashirov RV, Zemlyakov DI, Lyapkov AA, Kiselev SA. Kinetics of the metathesis polymerization of 5,6-di(methoxycarbonyl)bicycle[2.2.1]hept-2-enes on an original Hoveyda-Grabbs II type catalyst. Kinetics and Catalysis. 2013;**54**(4):494-499. DOI: 10.1134/S0023158413040010

[40] Delaude L, Demonceau A, Noels AF. Probing the Stereoselectivity of the ruthenium-catalyzed ring-opening metathesis polymerization of norbornene and norbornadiene diesters. Macromolecules. 2003;**36**(5):1446-1456. DOI: 10.1021/ma021315x

[41] Schrock RR. Synthesis of stereoregular ROMP polymers using molybdenum and tungsten imido alkylidene initiators. Dalton Transactions. 2011;**40**(29):7484-7495. DOI: 10.1039/c1dt10215j

[42] Flook MM, Ng VWL, Schrock RR. Synthesis of cis,syndiotactic ROMP polymers containing alternating enantiomers. Journal of the American Chemical Society. 2011;**133**(6):1784-1786. DOI: 10.1021/ja110949f

[43] Oskam JH, Schrock RR. Rotational isomers of Mo(VI) alkylidene complexes and Cis/trans polymer structure: Investigations in ring-opening metathesis polymerization. Journal of the American Chemical Society. 1993;**115**(25):11831-11845. DOI: 10.1021/ja00078a023

[44] O'Dell R, McConville DH, Hofmeister GE, Schrock RR. Polymerization of enantiomerically pure 2,3-dicarboalkoxynorbornadienes and 5,6-disubstituted norbornenes by well-characterized molybdenum ring-opening metathesis polymerization initiators. Direct determination of tacticity in cis, highly tactic and trans, highly tactic polymers. Journal of the American Chemical Society. 1994;**116**(8):3414-3423. DOI: 10.1021/ja00087a028

[45] Pollino JM, Stubbs LP, Weck M. Living ROMP of exo-norbornene esters possessing Pd II SCS pincer complexes or diaminopyridines. Macromolecules. 2003;**36**(7):2230-2234. DOI: 10.1021/ma025873n

[46] Ashirov RV, Zemlyakov D, Lyapkov AA, Kiselev S, Vervacke D. The relative reactivity of 2,3-dicarbomethoxy-5-norbornenes in metathesis polymerization using the original N-chelating ruthenium carbene complex. Journal of Applied Polymer Science. 2014;**131**(8): 401301-401307. DOI: 10.1002/app.40130

Section 2

Biomaterial Content Polymer Composites

Chapter 3

Molecularly Imprinted Polymers for Pharmaceutical Compounds: Synthetic Procedures and Analytical Applications

Lawrence Madikizela, Nikita Tavengwa and
Vusumzi Pakade

Additional information is available at the end of the chapter

http://dx.doi.org/10.5772/intechopen.71475

Abstract

In this chapter, the synthetic procedures for molecularly imprinted polymers (MIPs) for pharmaceutical compounds are discussed. Regardless of its limitations, such as production of irregular particles and loss of sample during processing (crushing and sieving), bulk polymerization has been widely used compared to say precipitation and suspension polymerization partly due to its simplicity in synthesis and robustness. A comparison of indomethacin removal from aqueous solution by MIP particles prepared using bulk polymerization to those obtained from suspension polymerization showed that the particles from the former exhibited higher adsorption capacity. Furthermore, the chapter explores the strengths and limitations relating the use of pharmaceutical compounds as uni-templates, multi-templates and dummy templates. Also, the analytical applications of MIPs are discussed in more details with particular focus on molecularly imprinted solid-phase extraction (MISPE) of pharmaceuticals from environmental samples. This application (MISPE) is currently the most exploited in literature as more pharmaceutical drugs find their way into environmental water bodies.

Keywords: molecularly imprinted polymer, polymerization process, pharmaceuticals, polymer composites, analytical applications

1. Introduction

The development of molecularly imprinted materials is an area of intense research ever since its introduction by Wulff and Sarhan in 1972 [1]. Generations of scientists have been intrigued by the binding phenomena involved in interactions that occur between natural molecular species such as antibodies and biological receptors. As such, over the years, numerous approaches have been used to mimic these interactions [2]. Molecular imprinting technology is today a viable synthetic approach to design robust molecular recognition materials able to mimic these natural phenomena. The main advantages of molecularly imprinted polymers (MIPs) are their high selectivity and affinity for the target molecule used in the imprinting procedure. Applications of molecularly imprinted materials have grown to include areas such as separation sciences and purification as extraction and chromatographic sorbents [3, 4], chemical sensors [5], catalysis [6], drug delivery [7] and biological antibodies and receptors system [8]. Among the polymeric materials developed, MIPs are also one of the most attractive materials for bioanalytical and biomedical applications [9]. Due to the high selectivity of MIPs towards pharmaceutical compounds, there are a number of companies that have gone to an extent of commercializing the MIP sorbent. These companies include Supelco in Belefonte (PA, USA), Biotage in Barcelona (Spain, Europe) and MIP Technologies in Lund (Sweden, Europe).

On the other hand, the occurrence of pharmaceutical compounds in the environment, particularly in surface water and wastewater, has also intrigued the scientific community. Pharmaceutical compounds are drugs that are used for the purpose of preventing, curing, treating disease and improving the health of their consumers [10]. To date, different classes of pharmaceuticals are known such as non-steroidal anti-inflammatory drugs (NSAIDs), antibiotics, antiretroviral drugs and steroid hormones. Pharmaceutical compounds enter the aquatic environment through various sources that include households, wastewater treatment plants (WWTPs), hospitals and industrial units [11]. Also, some pharmaceuticals are known to be transported into the environment through human excretion as metabolites and as unaltered parent compounds [12]. For example, the NSAIDs such as ibuprofen, naproxen, ketoprofen and diclofenac are eliminated from the human body with 10, 70, 80 and 10% of unaltered compounds, respectively [12]. The eliminated compounds are often swept into WWTPs. High number of research studies have demonstrated the inability of WWTPs to completely remove pharmaceuticals during the sewage treatment processes [13–15].

The formulation and the preparation of new pharmaceutical compounds conducted by companies can represent a danger for the environment because of their toxic potential. Despite concerns about potential risks associated with the presence of pharmaceuticals and personal care products in the environment, few toxicological data address the health and environmental effects of these compounds [16]. A recent review by Madikizela et al. [17] pointed out that there are indeed some traces of pharmaceutical compounds in water bodies even in many African counties that are least developed. Clofibric acid is regarded as one of the most persistent drug residues with an estimated persistence in the environment of 21 years, being frequently detected in environment monitoring of pharmaceuticals all around the world [18].

Conventional methods such as biodegradation, photo catalysis and advanced oxidation have been applied for the treatment of pharmaceutical contaminants [19]. Analytical tests are required for environmental monitoring of pharmaceutical drugs in order to evaluate the success of the treatment method. The complexity of the environmental samples has commanded the availability of selective analytical methods for the quantification of these

http://dx.doi.org/10.5772/intechopen.71475

pharmaceutical compounds. Therefore, MIPs are developed and used in sample preparation for the purpose of increasing selectivity of analytical method, and sensitivity when applied as a sorbent in the sample pre-concentration step. Other sorbents that have been used for the extraction of pharmaceuticals compounds include biochars, chitosan, silica, zeolites, graphene, clays and carbon [20].

Selectivity is an important parameter in analytical chemistry. However, many materials that are used for the extraction of pharmaceuticals such as hydrophilic lipophilic balance and C_{18} sorbents lack this feature. On the other hand, MIPs have long since known to be attractive in this regard. **Figure 1** [21] shows the chromatograms (red line) which illustrates the efficiency of the molecularly imprinted polymer—solid-phase extraction (MIP-SPE) procedure (extraction rate of about 90%) and the advantages of both the concentration and sample clean-up with very low background and no interferences close to the retention time of 17β-estradiol. It is also noteworthy to point out that not only selectivity is enhanced but MIPs have also the capacity to pre-concentrate pharmaceutical compounds. This is especially important as pharmaceuticals are detected and quantified in very low concentrations. This further implies that other ordinary cheap instruments can be used for environmental analysis beside the mass spectrometry (MS) detection that is known to have low detection limits and can detect pharmaceuticals at trace levels.

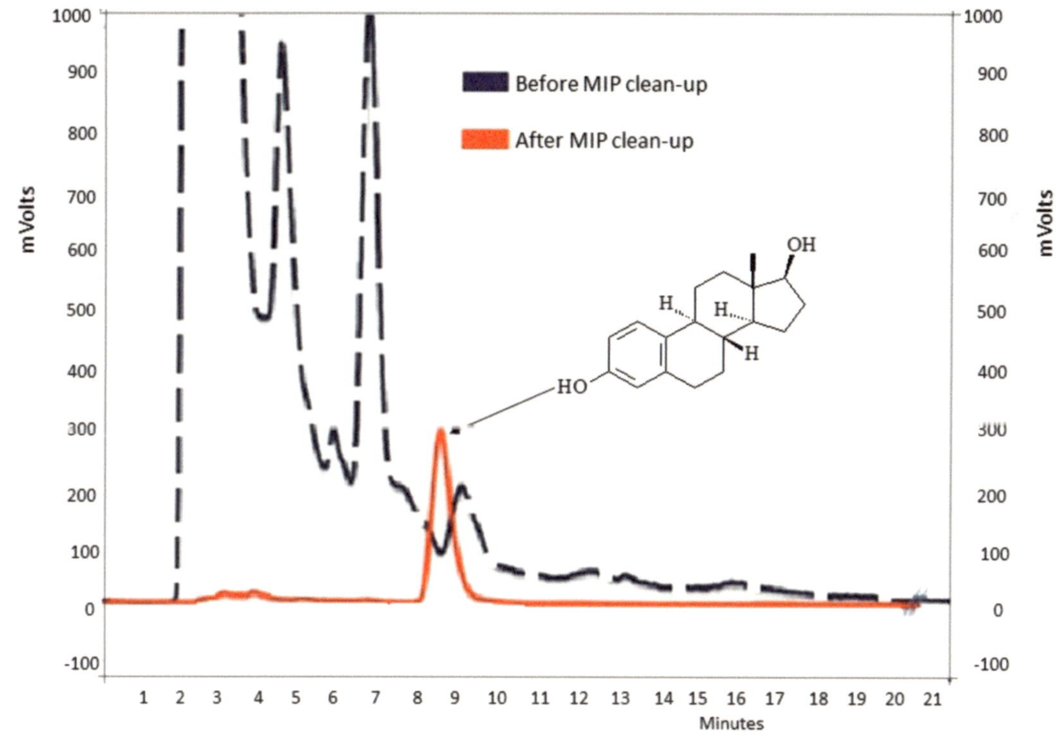

Figure 1. HPLC-FLD chromatograms obtained after extracts clean-up with MIP-SPE (AFFINIMIP® SPE Estrogens; Polyintell) of 100 mL of seine water spiked at 0.5 ng.mL^{-1} with 17β-estradiol (—) and before MIP clean-up (—) [21].

2. Synthetic approaches

The molecularly imprinted polymerization techniques and methods have been well described in literature [22]. Basically, the imprinting process involves self-assembled of selected functional monomer around a template molecule followed by polymerization in the presence of a cross-linker [23]. The template is then removed from the polymer matrix, thus, leaving behind a cavity complementary in functional group, size and shape, which is available to strongly bind compounds that are closely related to the template molecule. This is demonstrated in **Figure 2** using the synthesis of MIP for fluconazole, an antifungal agent, as an example [24]. In their synthetic reaction [24], they used methacrylic acid, ethylene glycol dimethacrylate and fluconazole as functional monomer, cross-linker and template, respectively. Despite its

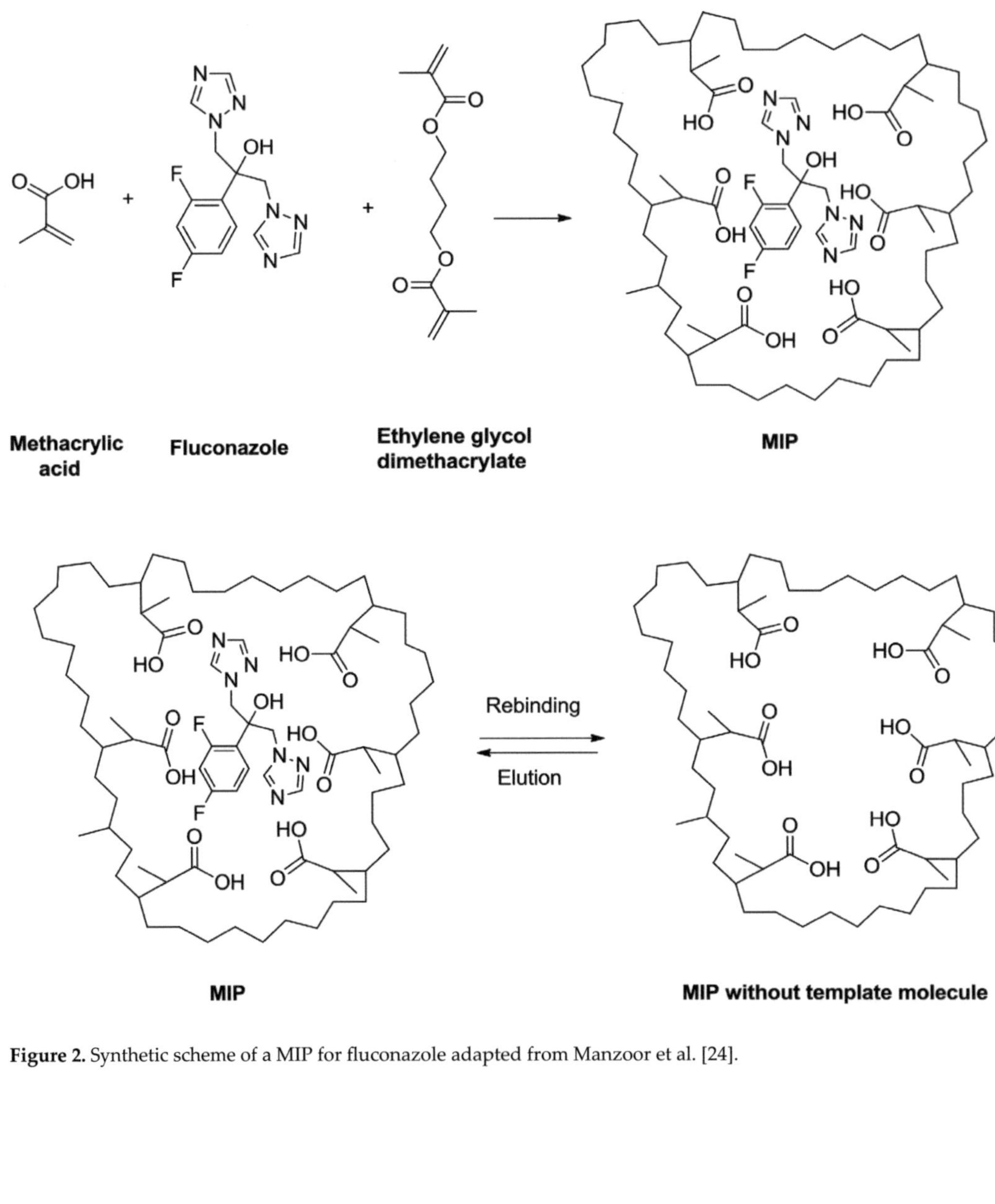

Figure 2. Synthetic scheme of a MIP for fluconazole adapted from Manzoor et al. [24].

disadvantages like wastage of material during processing (sieving and crushing), bulk polymerization remains the most common imprinting method for pharmaceutical compounds compared to other polymerization techniques such as precipitation, suspension and emulsion polymerizations [25]. Bulk polymerization has been practically proven to give higher binding capacity for indomethacin when compared to suspension polymerization [25]. In this instance, indomethacin binding capacities for MIP prepared by bulk and suspension polymerization were approximately 0.35 and 0.15 $mg.g^{-1}$, respectively.

3. Pharmaceutical compounds as templates

Templates are the most important reagent in the spatial arrangement of functional monomers during polymerization. Many pharmaceutical compounds of different therapeutic classes have been imprinted using the single-template and multi-template synthetic approaches [26–28]. Some of the imprinted compounds are given in **Table 1**. It has been observed that the non-steroidal anti-inflammatory drugs (NSAIDs) and analgesics with the exception of fenoprofen which are frequently detected in environmental waters have been imprinted [29]. NSAIDs are known as the most consumed pharmaceuticals and in most cases their MIPs have been applied as selective SPE sorbents [29]. Most pharmaceutical drugs have hydroxyl and carbonyl functional groups which makes the imprinting process easy by utilizing their ability to form hydrogen bonding with several functional monomers. Also, due to the presence of aromatic rings in some pharmaceutical compounds and functional monomers, it is possible to have electrostatic interactions between the rings of different molecules. In this aspect, Farrington and Regan [30] demonstrated computationally the electrostatic interaction between the aromatic rings of ibuprofen (template) and 4-vinyl pyridine (functional monomer). Currently, MIPs are developed using one pharmaceutical compound as the template (uni-templating and dummy-templating). Also, there are procedures that have been shown in literature where many pharmaceuticals are used as multi-templates [26, 28]. In multi-templating approach, equal amounts of templates are added simultaneously into the polymerization mixture. Once removed at the end polymerization reaction, the resulting MIP can selectively re-bind the removed compounds from the environmental samples.

3.1. Physico-chemical properties of pharmaceutical compounds

Few pharmaceutical compounds that have been used in molecular imprinting as template molecules are given in **Table 1**. It can be seen from their molecular structures that pharmaceutical compounds compose of a variety of functional groups. The presence of functional groups in the template molecule makes it easy to undergo the molecular interactions such as hydrogen bonding. For example, the presence of carboxylic group in ibuprofen allows for hydrogen bonding in acidic conditions with 2-vinyl pyridine functional monomer [31]. Such hydrogen bonds are easy to break by washing the MIP with acetic acid which is a small molecule that can penetrate the pores of the MIP, thereby, disrupting the molecular interactions. This leads to easy regeneration of the MIP.

Therapeutic class	Compound	Chemical structure
NSAID	Ibuprofen	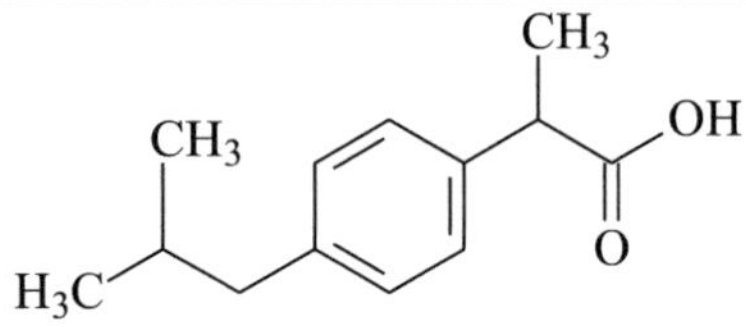
Antidiabetic	Metformin	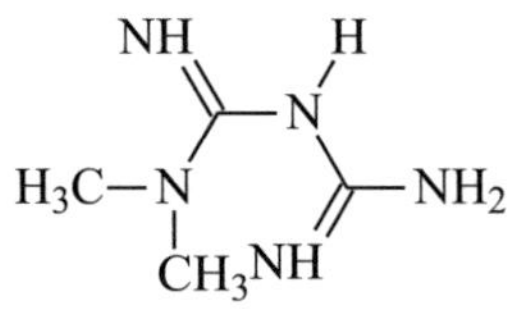
ARV	Efavirenz	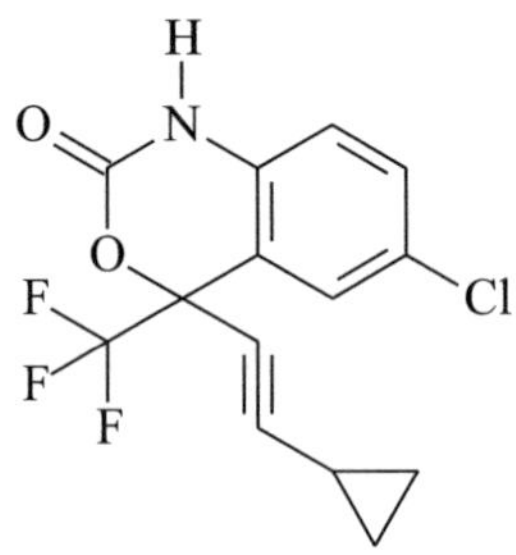
Antibiotic	Tetracycline	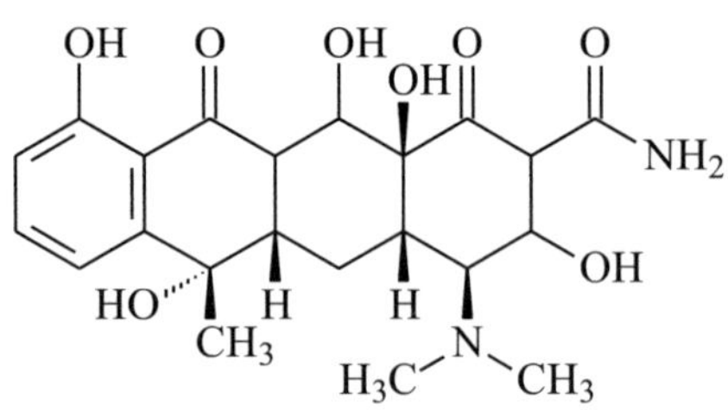
Anti-epileptic	Carbamazepine	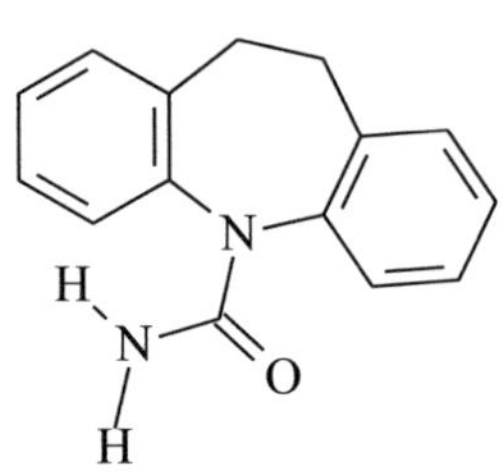

Table 1. Physico-chemical properties of selected imprinted pharmaceuticals.

The pK_a of a drug influences lipophilicity, solubility, protein binding and permeability which in turn directly affects pharmacokinetic (PK) characteristics such as absorption, distribution, metabolism and excretion [32, 33]. It is very important to understand the physico-chemical properties of templates and functional monomers prior to any MIP application. This is highlighted in the work reported by Dai et al. [34]. In their work, the adsorption efficiency of clofibric acid decreased significantly with the increase of pH when the pH was between 6 and 12.

This phenomenon could be explained by the ionization of clofibric acid (pK_a of 3.18) which could occur under strong basic condition. Therefore, clofibric acid was negatively charged. On the other hand, the functional monomer of 2-vinyl pyridine (pK_a of 4.98) used in the synthesis of MIP could also be negatively charged. It is known that the —COOH groups in the selective binding cavity of MIP play a key role in the rebinding of target compounds. The adsorption at basic pH could be due to the hydrophobic interactions [34].

3.2. Uni-templating

In most cases, uni-templating is done in order to extract one target analyte based on one attractive attribute such as bioactivity or ease excretion from human metabolism or the negative effect of that analyte such as toxicity or widespread in the environment. In the synthesis of MIPs, the target compounds are usually used as the template molecules. A diversity of pharmaceutical compounds have been detected in environmental waters [17, 35], therefore, most pharmaceuticals have been imprinted for the purpose of developing selective analytical methods. Conventionally, a single-template is imprinted which subsequently lead to the isolation of one pharmaceutical compound from water matrix [36]. Many pharmaceuticals that include ketoprofen [3], indomethacin [25], 17β-estradiol [27] and diclofenac [37] have been imprinted using the single templates. MIPs that have been synthesized using this approach usually possess high selectivity towards the compound that was used as template molecule. High selectivity maybe due to molecular recognition that could be strongly influenced by functional groups, shape and size of the target compound. In this case, selectivity is usually evaluated by extracting a mixture of organic compounds that consist of the target compound and other structurally related compounds. Zunngu et al. [3] tested selectivity of ketoprofen MIP for its ability to extract similar compounds (triclosan, gemfibrozil and fenoprofen) along the target compound from spiked deionized water. Their results showed accepted recovery (104%) for ketoprofen, and for competitors the recoveries did not exceed 20%. In a different study, Ming et al. [27] demonstrated the competitive adsorption ability of 17β-estradiol MIP towards the target compound in the presence of estriol, estrone, bisphenol A and hexestrol in aqueous solutions. Single-template MIPs usually lead to superior sample clean-up that subsequently results in cleaner chromatograms, however, this approach do not allow for a simultaneous multi-compound analysis. Multi-compound analysis can only be performed using uni-templating procedure after physical mixing a number of MIP for individual compounds. However, this approach is not financial feasible as it is expensive to synthesize a number of MIPs whose mixtures will be used to target a number of pharmaceuticals.

3.3. Multi-templating

One of the advantages of MIPs is the selective extraction of analytes in complex matrices. However, at times it is desired that a class/many compounds are removed or extracted from the environmental or real samples simultaneously. To achieve this, researchers have explored the possibility of imprinting multiple templates all at once for the

Dummy template	Target compounds		Selectivity	Recovery (%)	Reference
Diphenylamine 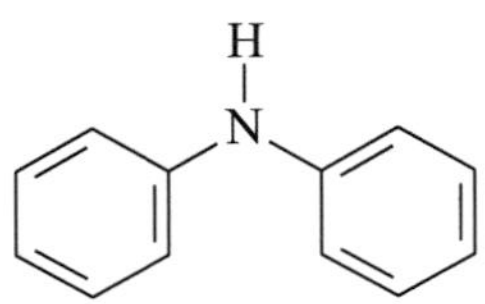	Diclofenac 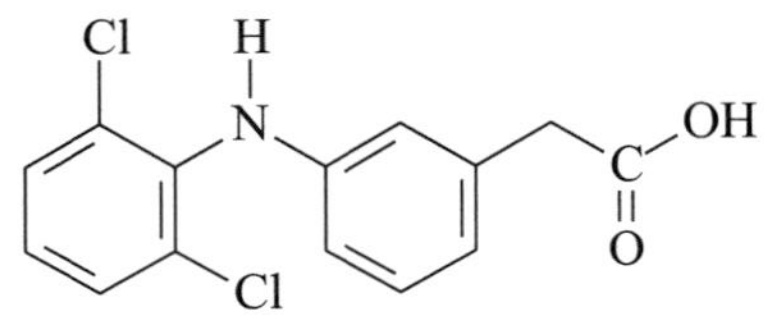	Mefenamic acid 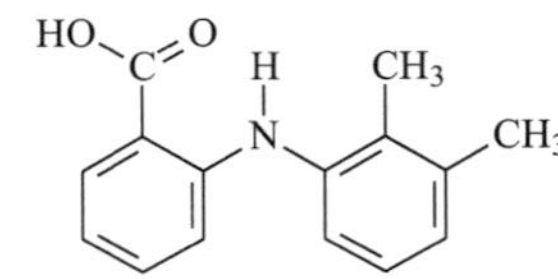	Selective in the presence of ibuprofen, naproxen, ketoprofen and salicyclic acid	100–112	[44]
Phenothiazine 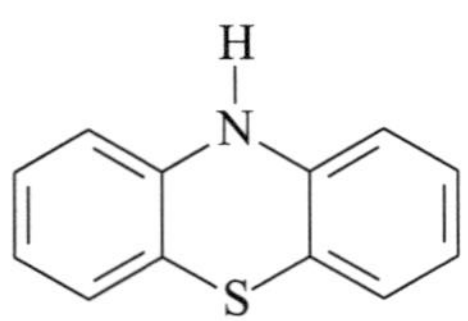	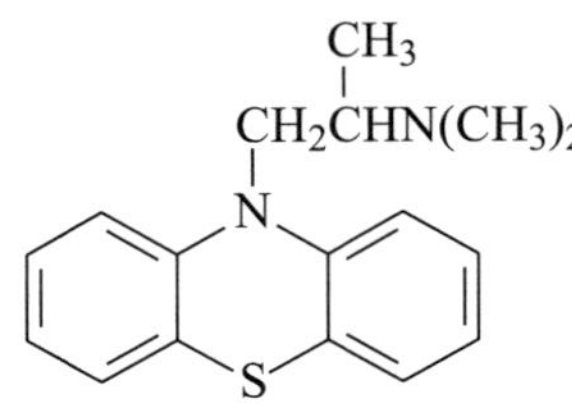	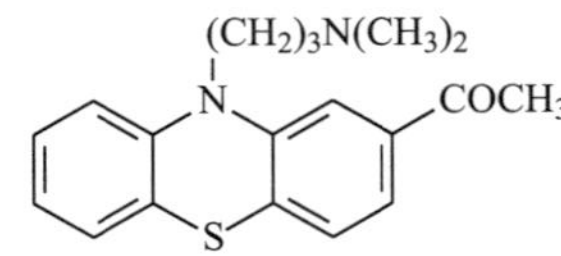 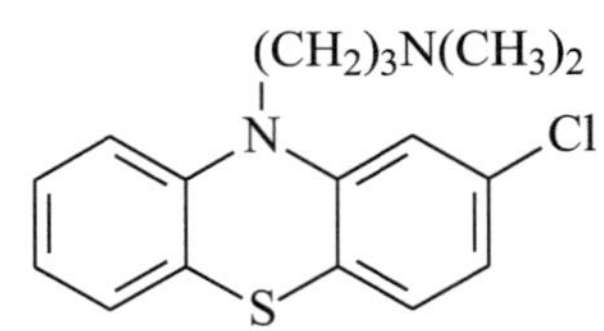	Selective in the presence of tetracycline, naphthalene and anthracine	93–98	[42]
2-chlorophenothiazine 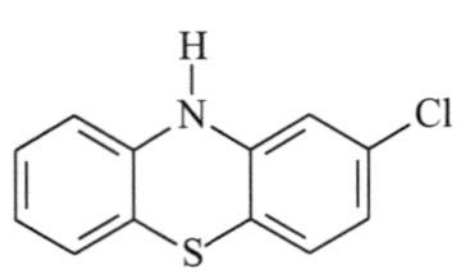	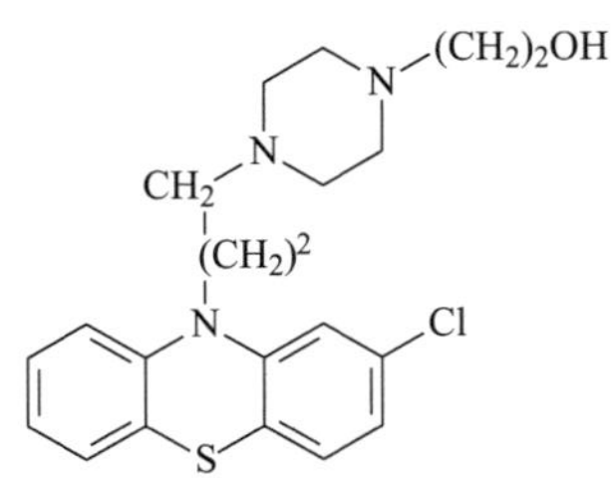				
Salbutamol 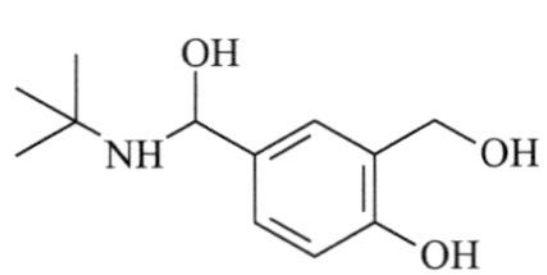	Ractopamine 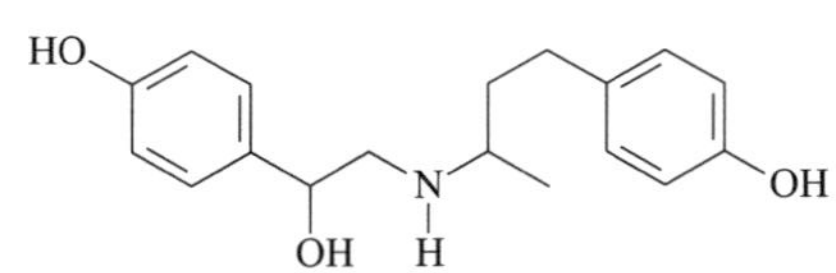		Selectivity was done chromatographically using clenbuterol, terbutaline and adrenaline as competitors.	—	[41]

Dummy template	Target compounds		Selectivity	Recovery (%)	Reference
Daidzein 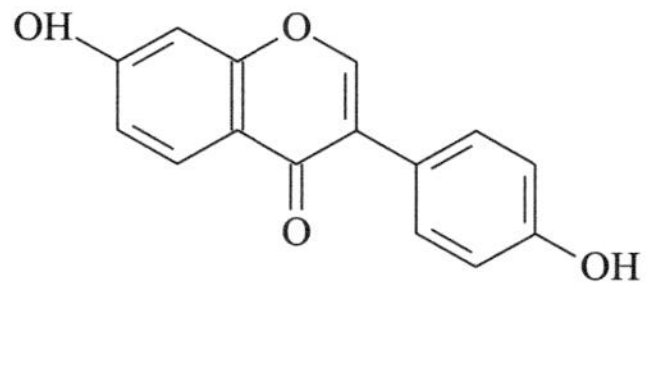	Fluoroquinolones*(1) Norfloxacin 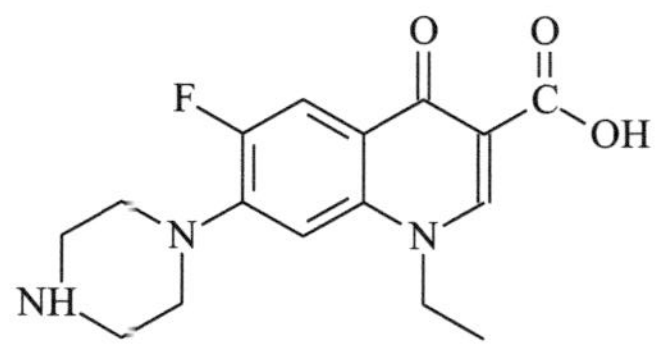	(2) Gatifloxacin 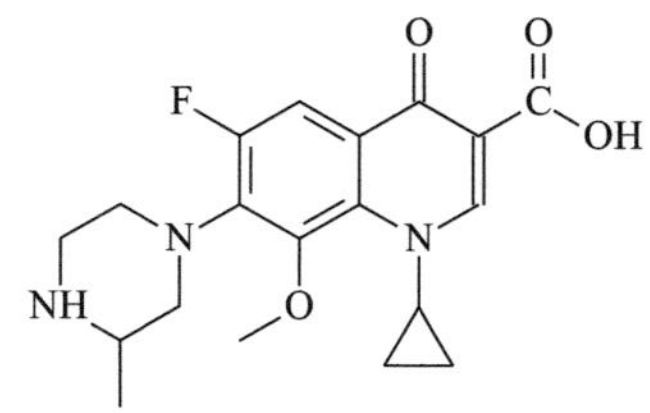		64–103	[43]

*Fluoroquinolones—only two chemical structures are given, however, there were eight target compounds (fleroxacin, ofloxacin, norfloxacin, pefloxacin, ciprofloxacin, lomefloxacin, enrofloxacin and gatifloxacin).

Table 2. Chemical structures of dummy template molecules, their relative extracted compounds and analytical performance.

pre-concentration and extraction of a certain group of compounds where a cocktail of pharmaceutical compounds is used in the polymerization set-up. Practical examples for this include the work of Duan et al. [26] where five acidic pharmaceuticals which are ibuprofen, naproxen, ketoprofen, diclofenac and clofibric acid, were used as multi-templates in a synthesized MIP that showed selective recognition and ability to extract these target compounds from lake water and WWTP effluent using molecularly imprinted solid-phase extraction (MISPE) technique. In their work, Duan et al. [26] obtained the recoveries that were greater than 95% for all five acidic pharmaceuticals in lake water and wastewater spiked with 1 $\mu g.L^{-1}$ of each compound. Dai et al. [38] have also reported the selective removal of the same group of pharmaceutical compounds using a multi-template MIP from contaminated water. These researchers demonstrated the selectivity of the multi-template MIP in the presence of fenoprofen and carbamazepine (both pharmaceutical compounds). In their study, the removal efficiency for the five target pharmaceuticals in water was greater than 80%, whereas, less than 40% was reported for fenoprofen and carbamazepine used as competitors. In the same aspect, a dual template (naproxen and ketoprofen) MIP has been reported [28], where the ability to recognize the template molecules was tested chromatographically in the presence of structural analogues (ibuprofen, fenbufen, fenoprofen and flurbiprofen). Also, Dai et al. [34] prepared a novel double-template MIP by precipitation polymerization using carbamazepine and clofibric acid as the double templates.

For a multi-template (naproxen, ibuprofen and diclofenac) MIP, it was observed that the selectivity collapses easily during the extraction of target compounds from aqueous phase [31]. This could be strongly influenced by bigger cavities that are created due to the usage of multi-templates. This could allow the easy access of many presumably smaller molecules into the binding sites. It has been demonstrated that the untargeted compounds can be selectively washed off from the multi-template MIP surface due to their weaker non-specific interactions with the polymer [31]. Contrarily, some researchers have indicated that the use of a dummy template during polymerization increased the selectivity of the final MIPs that target more that target one compound [39, 40].

3.4. Dummy-templating

The usage of target compounds as template molecules could have negative impact on the analysis of real samples due to their bleeding upon application into the sample matrix. This could be severe in the cases of incomplete template removal. To avoid this problem, the use of dummy templates for the synthesis of MIPs has been proposed in many studies [41–43]. In certain instances, the selected dummy templates exhibit the properties of more than one compound, and its chemical structure is closely related or similar to more than one pharmaceutical drug [42, 44]. In such cases, the prepared MIP is able to selectively extract more than one compound. This has been demonstrated by the synthesis of a MIP using diphenylamine as the template whose chemical structure closely resembles that of both diclofenac and mefenamic acid (**Table 2**) [44]. As shown in **Table 2**, the dummy molecule (diphenylamine) for both diclofenac and mefenamic acid can be characterized by two phenyl groups which are both attached to the amine group, such groups also appear on the structures of the target compounds. Similarly, the approach has also been reported in the synthesis of MIP required for SPE of phenothiazines from meat samples, where phenothiazine and 2-chlorophenothiazine were used as dummy templates, thereby taking the advantage of their core chemical structures that compose of phenothiazine [42]. Both MIPs synthesized with phenothiazine and 2-chlorophenothiazine dummy templates were able to capture four different phenothiazines (acepromazine, promethazine, perphenazine and chlorpromazine), simultaneously. In a different work, a compound, daidzein, was used as a non-poisonous dummy template in the synthesis of MIP for fluoroquinolone antibiotics [43]. Their synthesized MIP was successfully applied for matrix solid-phase dispersion extraction of eight fluoroquinolones from fish samples. Such work gave high recoveries and selectivity for target compounds (**Table 2**). Similarly, the analysis of single pharmaceutical drug, ractopamine, commonly used for the treatment of asthma has been performed using dummy template MIP which has been synthesized using salbutamol as a dummy template [41]. The drawback that could be associated with the use of dummy templates in molecular imprinting could rise during the applications in real samples. Selectivity can be reduced greatly due to the differences in the physico-chemical properties of the dummy template and the targeted compound(s). For instance, in **Table 2**, it can be seen that diphenylamine (dummy template) is relatively smaller, in terms of size, as compared to the diclofenac (target compound). The same can be observed in the case of phenothiazine presented in **Table 2**. This could lead to easy binding of smaller molecules with similar functional groups into MIPs.

4. Synthesis of polymer composites

Traditional imprinting produce particles with adsorption sites embedded deep into the polymer matrix resulting in difficulties with mass transfer of analyte molecules [45]. Dummy templating has been used to address the problem of elution of template molecules during MIP application in real world samples. This problem occurs due to incomplete template removal caused by molecules occupying deeper pores in the polymer matrix. However, as discussed earlier, the recognition sites created by the dummy might not perfectly fit the target analyte leading to inferior extraction efficiencies as when the template was used. Surface imprinting (polymer composite) has been proposed as an alternative to dummy templating for addressing the MIP template bleeding problem. Various materials including graphene oxide [46, 47], magnetic cross-linked chitosan [48, 49], silica [49, 50], carbon nanotubes [51], quantum dots [52] and nanoparticles [53, 54] have been used as scaffolds in the preparation of imprinted polymer composites for pharmaceutical compounds. The template is imprinted at or near the proximity of the surface, therefore, there are no deeper lying cavities that may cause slow release of template molecules and subsequently bleeding. The materials used as anchors usually have large surface area to volume ratio resulting in lower mass transfer resistance and faster rebinding due to accessibility of the binding site [47]. Cl-TiO_2 imprinted photo catalyst exhibited higher photo degradation rate (72%) of tetracycline under visible irradiation than non-imprinted photo catalyst [53]. Molecular imprinting magnetic γ-Fe_2O_3/cross-linked chitosan composites prepared by a microemulsion process were applied for the adsorption and degradation of norfloxacin (NOR) [48]. The MIP showed superior adsorption of NOR than its non-imprinted counterpart [48] and excellent selectivity of NOR adsorption in comparison to sulfadiazine, ofloxacin and phenol [49]. Elsewhere, an electrochemical sensor constructed by grafting MIP to multi-walled carbon nanotube (MWCNTs) surface immobilized on glass carbon electrode was evaluated for the determination of ceftazidime from human serum [51]. The functionalized MWCNTs played two roles, increasing the conductivity of a sensor and the amount of binding sites. The sensor demonstrated good precision, stability, sensitivity and selectivity for the target analyte. Recent work showed the synthesis of polymer composites for pharmaceutical drugs, where such materials are prepared for the purpose of sensory applications [52]. Prasad et al. [52] synthesized a novel monomeric graphene quantum dots—MIP-based nanocomposite directly at the surface of screen printed carbon electrode and applied for electrochemical detection of an anticancerous drug ifosfamide in biological and pharmaceutical samples. Their sensor gave the detection limit of 0.11 $ng.mL^{-1}$ (S/N = 3), without any matrix effect, cross-reactivity and false-positives.

5. Reusability

MIPs sorbents are easily regenerated by washing with organic solvents to remove the adsorbed compounds. Therefore, they can be applied for various extractions repeatedly. In this aspect, acetylsalicylic acid MIP successfully adsorbed (removal efficiency 75–78%) the target compound from acidic aqueous solutions six times [55]. After desorption, MIP was regenerated with a mixture of methanol and acetic acid (7:3, v/v) followed by methanol. There are numerous examples of reusability in literature that include those cited in **Table 3** for the MIPs prepared for selective extraction of acetylsalicylic acid [55], diclofenac [36] and multi-templates (ibuprofen, naproxen, ketoprofen, diclofenac and clofibric acid) [26]. Such work, clearly demonstrate that MIP can be reused many times without losing its adsorption capacity. This is an excellent advantage for MIP as it is a common knowledge that many adsorbents are discarded after a single use.

Template/target compound	Regeneration solvent	Successful applications	Reference
Acetylsalicylic acid	Methanol/acetic acid (7:3, v/v) and methnol	6 repeatable adsorption/desorption experiments with 75–78% removal efficiency	[55]
Diclofenac	Methanol/acetic acid (9:1, v:v)	30 binding/regeneration cycles with ≥95% recovery	[36]
Multi-templates*	Methanol/acetic acid (9:1, v/v)	20 adsorption and desorption cycles gave constant recovery (≥95%)	[26]

*Multi-templates were ibuprofen, naproxen, ketoprofen, diclofenac and clofibric acid.

Table 3. Examples for reusability of MIPs.

6. Analytical applications

In relation with pharmaceutical compounds, MIPs are used in analytical applications such as sample preparation as SPE sorbents [37], chromatographic stationary phase [56] and biological sensing [51]. The analytical applications of MIPs are very useful for various reasons, such as they provide higher selectivity than the conventional sorbents, they also reduce the matrix influence on the resulting chromatograms and they lead to high sample enrichment factors [57]. Besides the use of MIPs in analytical applications such as sample preparation, they are widely evaluated as selective adsorbents in contaminated water. MIPs are introduced as clean-up adsorbents in environmental waters for removal of pharmaceuticals (batch adsorption) [31, 55].

6.1. Sample preparation

Sample preparation for the determination of pharmaceuticals in aqueous samples is, in most cases, preceded by a filtration method prior to pre-concentration [58]. However, this step may lead to loss of some compounds bound to particulate matter. Typical example to this, is the detection of trace levels of mefenamic acid (a hydrophobic compound) at $\mu g.kg^{-1}$ on the suspended solids following filtration [59]. Therefore, removal efficiencies and mass loadings may be affected by the filtration step [58]. Solid-phase extraction (SPE) techniques, on the other hand, have shown fairly good pre-concentration and extraction efficiencies for hydrophobic compounds. However, commercial reversed phase-based adsorbents used in SPE have not shown satisfactory efficiencies for the pre-concentration of polar organic compounds. It has been suggested that the molecular recognition brought about by MIPs could address this downfall of SPE [58]. Instead of conventional sorbents, the SPE cartridges are in this case packed with MIP particles. That is, the synthesized MIP particles are slurry or dry packed in between two frits inside the solid-phase extraction cartridge [60, 61], referred to as MISPE. After packing, the cartridge is then conditioned prior to the loading of sample solutions. Thereafter, the MISPE cartridge is washed for removal of sample interferences and the target compounds are eluted with a suitable organic solvent. MISPE has been widely used for selective extraction of pharmaceuticals from various matrices that include plasma, urine and water samples [25, 57, 62]. In most cases, MISPE is applied where target compounds are extracted from solution into the solid material.

In addition, other mode of solid-phase extraction for ARV drug (abacavir) such as solid-phase microextraction (SPME) has been reported [63]. The molecularly imprinted SPME technology for drug analysis has been described in great details by Ansari and Karimi [22]. These authors focussed on the progress, challenges and trends in trace determination of different drugs.

6.1.1. Molecularly imprinted solid-phase extraction

Pharmaceuticals from different classes have been extracted and pre-concentrated using MISPE (**Table 4**). Most common non-steroidal anti-inflammatory drugs (NSAIDs) with the exception of fenoprofen have been imprinted and their MIPs were applied in the form of MISPE from environmental samples [29]. This is expected as NSAIDs are classified as the most consumed pharmaceuticals by humans with antipyretic activities in some countries such as South Africa [64]. The emerging environmental pollutants such as antiretroviral drugs (ARVs) are imprinted [63, 65], however, there is still limited/no information on their environmental extraction using MISPE. Most MISPE applications for ARVs are based on their extraction from biological samples [66, 67]. The application of MISPE allows for pre-concentration of various analytes from environmental samples which in turn lead to very low detection limits in $\mu g.L^{-1}$ to $ng.L^{-1}$ levels (**Table 4**). Based on higher extraction efficiency or percent recoveries for pharmaceuticals, MIPs show strong ability to extract such drugs from complex sample matrices such as wastewater. As can be seen in **Table 4**, various amounts of MIPs are used in SPE. Small quantities as demonstrated by Zunngu et al. [3] are the significant of potential application in miniaturization techniques.

6.2. Chromatographic analysis

One of the most important applications of MIPs is their usage as the chromatographic stationary phases. This is done by slurry packing the prepared MIP into the stainless still chromatographic column. During the application, the imprinted molecule binds strongly to the packing material, which results in its strong retention and longer retention time [56]. This application was demonstrated in literature where a chiral stationary phase for the enantioselective separation of naproxen was reported [56]. In their work [56], a MIP was synthesized using (*S*)-naproxen as the template and evaluated for chromatographic separation. Racemic naproxen was efficiently resolved on the MIP with (*S*)-naproxen eluted last. Similarly, Haginaka and Sanbe, [70], synthesized a uniformly sized MIP for (*S*)-naproxen that gave good enantioselectivity and resolution for naproxen. In addition, uniform-sized MIP material for (*S*)-propranolol when applied as chromatographic stationary phase has shown the ability to separate (*S*)-propranolol from a mixture that contains some structurally related β-adrenergic antagonists [71]. Due to the strong binding of target compound onto MIP, peak tailing on the chromatogram is usually evident. Therefore, there are opportunities relating to the improvement of the quality of the resulting chromatograms.

Analyte	Sorbent amount (mg)	Environmental sample loading	Elution	Analytical method and detection limit	Reference
Diclofenac	35	1000 mL wastewater and river water	2 mL of methanol/ acetic acid (9:1, v:v)	LC–MS/MS LOD – Not reported	[36]
Carbamazepine	200	100 mL wastewater effluent at pH 11	5 mL of methanol	LC-UV LOD – 25 $\mu g.L^{-1}$	[68]
Metformin	50	50 mL aqueous samples including wastewater, pH 10	1 mL of acetic acid and methanol (1:9)	LC-DAD-ESI/MS LOD – 1.5–3.4 $ng.L^{-1}$	[69]
Ketoprofen	14	50 mL wastewater at pH 5	1 mL of methanol	LC-UV LOD – 0.23 $\mu g.L^{-1}$	[3]
Indomethacin	200	100 mL river water at pH 5	2 mL methanol	LC-UV LOD – 0.03 $\mu g.mL^{-1}$	[25]

Table 4. MISPE of pharmaceuticals from environmental waters.

7. Results and suggestions

In the earlier sections of this chapter, the evidence of the improved selectivity due to the application of MIPs in analytical methods has been demonstrated (**Figure 1**). In numerous occasions, the quantification of some pharmaceutical drugs in the environment has been performed after sample clean-up and analyte(s) pre-concentration using MISPE. After pre-concentrated with MISPE, pharmaceutical compounds have been detected at concentration levels that range from low $ng.L^{-1}$ to $\mu g.L^{-1}$ [3, 26, 37, 44, 61]. In this chapter, it was further elaborated that there are MIP sorbents that are available commercially, and therefore, it is suggested that more sorbents should be available in the near future as more pharmaceuticals are being detected in the environment. Moreover, there are no reports in literature for MIPs synthesized for a number of pharmaceutical compounds more especially those that are new in the market. For example, in their recent review paper, Madikizela et al. [29] observed that the MIP for fenoprofen is yet to be developed. In different perspective, the potential for MIPs to be applied as chromatographic stationary phases for separation of complex mixtures such as enantiomers have been investigated extensively in the early 2000 [56, 70, 71]. Due to the promising results reported in previous years, it is suggested that this area should be exploited more carefully in order to improve the quality of chromatographic peaks, that could lead to better quality of analysis and results that are more reliable can be achieved. This of cause could lead to the introduction of new stationary phases by the manufacturers of chromatographic equipment and consumables.

Author details

Lawrence Madikizela[1]*, Nikita Tavengwa[2] and Vusumzi Pakade[3]

*Address all correspondence to: lawremcem2@dut.ac.za

1 Department of Chemistry, Durban University of Technology, Durban, South Africa

2 Department of Chemistry, University of Venda, Thohoyandou, South Africa

3 Department of Chemistry, Vaal University of Technology, Vanderbjlpark, South Africa

References

[1] Wulff G, Sarhan A. Über die Anwendung von enzymanalog gebauten Polymeren zur Racemattrennung. Angewandte Chemie. 1972;**84**:64

[2] Puoci F, Lemma F, Cirillo G, Picci N, Matricardi P, Alhaique F. Molecularly imprinted polymers for 5-fluorouracil release in biological fluids. Molecules. 2007;**12**:805-814

[3] Zunngu SS, Madikizela LM, Chimuka L, Mdluli PS. Synthesis and application of a molecularly imprinted polymer in the solid-phase extraction of ketoprofen from wastewater. Comptes Rendus Chimie. 2017;**20**:585-591

[4] Chen D, Fu Q, Du W, Sun S, Huang P, Chang C. Preparation and evaluation of monolithic molecularly imprinted stationary phase for S-naproxen. Journal of Pharmaceutical Analysis. 2011;**1**:26-31

[5] Cieplak M, Szwabinska K, Sosnowska M, Chandra BKC, Borowicz P, Noworyta K, D'Souza F, Kutner W. Selective electrochemical sensing of human serum albumin by semi-covalent molecular imprinting. Biosensors and Bioelectronics. 2015;**74**:960-966

[6] Kirsch N, Hedin-Dahlström J, Henschel H, Whitcombe MJ, Wikman S, Nicholls IA. Molecularly imprinted polymer catalysis of a Diels-Alder reaction. Journal of Molecular Catalysis B: Enzymatic. 2009;**58**:110-117

[7] Cirillo G, Parisi OI, Curcio M, Puoci F, Iemma F, Spizzirri UG, Picc N. Molecularly imprinted polymers as drug delivery systems for the sustained release of glycyrrhizic acid. Journal of Pharmacy and Pharmacology. 2010;**62**:577-582

[8] Panagiotopoulou M, Kunath S, Medina-Rangel PX, Haupt K, Bui BTS. Fluorescent molecularly imprinted polymers as plastic antibodies for selective labeling and imaging of hyaluronan and sialic acid on fixed and living cells. Biosensors and Bioelectronics. 2017;**88**:85-93

[9] Lee SH, Doong RA. Design of size-tunable molecularly imprinted polymer for selective adsorption of pharmaceuticals and biomolecules. Biosensors and Bioelectronics. 2016;**7**:228

[10] Jelic A, Gros M, Ginebreda A, Cespedes-Sanchez R, Ventura F, Petrovic M, Barcelo D. Occurrence, partition and removal of pharmaceuticals in sewage water and sludge during wastewater treatment. Water Research. 2011;**45**:1165-1176

[11] Kostopoulou M, Nikolaou A. Analytical problems and the need for sample preparation in the determination of pharmaceuticals and their metabolites in aqueous environmental matrices. Trends in Analytical Chemistry. 2008;**27**:1023-1035

[12] Kasprzyk-Hordern B, Dinsdale RM, Guwy AJ. Illicit drugs and pharmaceuticals in the environment – Forensic applications of environmental data. Part 1: Estimation of the usage of drugs in local communities. Environmental Pollution. 2009;**157**:1773-1777

[13] Behera SK, Kim HW, Oh J, Park H. Occurrence and removal of antibiotics, hormones and several other pharmaceuticals in wastewater treatment plants of the largest industrial city of Korea. Science of the Total Environment. 2011;**409**:4351-4360

[14] Gracia-Lor E, Sancho JV, Serrano R, Hernandez F. Occurrence and removal of pharmaceuticals in wastewater treatment plants at the Spanish Mediterranean area of Valencia. Chemosphere. 2012;**87**:453-462

[15] Liu H, Lam JCW, Li W, Yu H, Lam PKS. Spatial distribution and removal performance of pharmaceuticals in municipal wastewater treatment plants in China. Science of the Total Environment. 2017;**586**:1162-1169

[16] Dussault EB, Balakrishnan VK, Sverko E, Solomon KR, Sibley PK. Toxicity of human pharmaceuticals and personal care products to benthic invertebrates. Environmental Toxicology and Chemistry. 2008;**27**:425-432

[17] Madikizela LM, Tavengwa NT, Chimuka L. Status of pharmaceuticals in African water bodies: Occurrence, removal and analytical methods. Journal of Environmental Management. 2017;**193**:211-220

[18] Heberer T. Occurrence, fate, and removal of pharmaceutical residues in the aquatic environment: A review of recent research data. Toxicology Letters. 2002;**131**:5-17

[19] Kümmerer K. Antibiotics in the aquatic environment — A review — Part I. Chemosphere. 2009;**75**:417-434

[20] Kyzas GZ, Fu J, Lazaridis NK, Bikiaris DN, Matis KA. New approaches on the removal of pharmaceuticals from wastewaters with adsorbent materials. Journal of Molecular Liquids. 2015;**209**:87-93

[21] Lucci P, Pacetti D, Nunez O, Frega NG. Current Trends in Sample Treatment Techniques for Environmental and Food Analysis. Chapter 5. InTech; 2012. p. 127-164

[22] Ansari S, Karimi M. Recent progress, challenges and trends in trace determination of drug analysis using molecularly imprinted solid-phase microextraction technology. Talanta. 2017;**164**:612-625

[23] Gama MR, Bottoli CBG. Molecularly imprinted polymers for bioanalytical sample preparation. Journal of Chromatography B. 2017;**1043**:107-121

[24] Manzoor S, Buffon R, Rossi AV. Molecularly imprinted solid phase extraction of fluconazole from pharmaceutical formulations. Talanta. 2015;**134**:1-7

[25] Yang T, Li Y, Wei S, Li Y, Deng A. Development of a selective molecularly imprinted polymer-based solid-phase extraction for indomethacin from water samples. Analytical and Bioanalytical Chemistry. 2008;**391**:2905-2914

[26] Duan Y, Dai C, Zhang Y, Chen L. Selective trace enrichment of acidic pharmaceuticals in real water and sediment samples based on solid-phase extraction using multi-templates molecularly imprinted polymers. Analytica Chimica Acta. 2013;**758**:93-100

[27] Ming W, Wang X, Lu W, Zhang Z, Song X, Li J, Chen L. Magnetic molecularly imprinted polymers for the fluorescent detection of trace 17β-estradiol in environmental water. Sensors and Actuators B: Chemical. 2017;**238**:1309-1315

[28] Zhang J, Li F, Wang X, Xu D, Huang Y, Liu Z. Preparation and characterization of dual-template molecularly imprinted monolith with metal ion as pivot. European Polymer Journal. 2016;**80**:134-144

[29] Madikizela LM, Tavengwa NT, Chimuka L. Applications of molecularly imprinted polymers for solid-phase extraction of non-steroidal anti-inflammatory drugs and analgesics from environmental waters and biological samples. Journal of Pharmaceutical and Biomedical Analysis. 2018;**147**:624-633

[30] Farrington K, Regan F. Investigation of the nature of MIP recognition: The development and characterisation of a MIP for ibuprofen. Biosensors and Bioelectronics. 2007;**22**:1138-1146

[31] Madikizela LM, Chimuka L. Synthesis, adsorption and selectivity studies of a polymer imprinted with naproxen, ibuprofen and diclofenac. Journal of Environmental Chemical Engineering. 2016;**4**:4029-4037

[32] Avdeef A. Physicochemical profiling (solubility, permeability and charge state). Current Topics in Medicinal Chemistry. 2001;**1**:277-351

[33] Kerns EH, Di L. Physicochemical profiling: Overview of the screens. Drug Discovery Today: Technologies. 2004;**1**:343-348

[34] Dai C, Zhang J, Zhang Y, Zhou X, Duan Y, Liu S. Removal of carbamazepine and clofibric acid from water using double templates–molecularly imprinted polymers. Environmental Science and Pollution Research. 2013;**20**:5492-5501

[35] Sun J, Luo Q, Wang D, Wang Z. Occurrences of pharmaceuticals in drinking water sources of major river watersheds, China. Ecotoxicology and Environmental Safety. 2015;**117**:132-140

[36] Dai C, Zhou X, Zhang Y, Liu S, Zhang J. Synthesis by precipitation polymerization of molecularly imprinted polymer for the selective extraction of diclofenac from water samples. Journal of Hazardous Materials. 2011;**198**:175-181

[37] Sun Z, Schussler W, Sengl M, Niessner R, Knopp D. Selective trace analysis of diclofenac in surface and wastewater samples using solid-phase extraction with a new molecularly imprinted polymer. Analytica Chimica Acta. 2008;**620**:73-81

[38] Dai C, Zhang J, Zhang Y, Zhou X, Duan Y, Liu S. Selective removal of acidic pharmaceuticals from contaminated lake water using multi-templates molecularly imprinted polymer. Chemical Engineering Journal. 2012;**211-212**:302-309

[39] Zhan W, Wei F, Xu G, Cai Z, Du S, Zhou X, Li F, Hu Q. Highly selective stir bar coated with dummy molecularly imprinted polymers for trace analysis of bisphenol A in milk. Journal of Separation Science. 2012;**35**:1036-1043

[40] Zhu R, Zhao W, Zhai M, Wei F, Cai Z, Sheng N, Hu Q. Molecularly imprinted layer-coated silica nanoparticles for selective solid-phase extraction of bisphenol A from chemical cleansing and cosmetics samples. Analytica Chimica Acta. 2010;**658**:209-216

[41] Du W, Fu Q, Zhao G, Huang P, Jiao Y, Wu H, Luo Z, Chang C. Dummy-template molecularly imprinted solid phase extraction for selective analysis of ractopamine in pork. Food Chemistry. 2013;**139**:24-30

[42] Song YP, Li N, Zhang HC, Wang GN, Liu JX, Liu J, Wang JP. Dummy template molecularly imprinted polymer for solid phase extraction of phenothiazines in meat based on computational simulation. Food Chemistry. 2017;**233**:422-428

[43] Sun X, Wang J, Li Y, Yang J, Jin J, Shah SM, Chen J. Novel dummy molecularly imprinted polymers for matrix solid-phase dispersion extraction of eight fluoroquinolones from fish samples. Journal of Chromatography A. 2014;**1359**:1-7

[44] Manzo V, Ulisse K, Rodríguez I, Pereira E, Richter P. A molecularly imprinted polymer as the sorptive phase immobilized in a rotating disk extraction device for the determination of diclofenac and mefenamic acid in wastewater. Analytica Chimica Acta. 2015;**889**:130-137

[45] Alexiadou DK, Maragou NC, Thomaidis NS, Theodoridis GA, Koupparis MA. Molecularly imprinted polymers for bisphenol A for HPLC and SPE from water and milk. Journal of Separation Science. 2008;**31**:2272-2282

[46] Wen T, Xue C, Li Y, Wang Y, Wang R, Hong J, Zhou X, Jiang H. Reduced graphene oxide-platinum nanoparticles composites based imprinting sensor for sensitively electrochemical analysis of 17β-estradiol. Journal of Electroanalytical Chemistry. 2012;**682**:121-127

[47] Roy E, Patra S, Tiwari A, Madhurin R, Sharma PK. Introduction of selectivity and specificity to graphene using an inimitable combination of molecular imprinting and nanotechnology. Biosensors and Bioelectronics. 2017;**89**:234-248

[48] Huang M, Zhou T, Wu X, Mao J. Adsorption and degradation of norfloxacin by a novel molecular imprinting magnetic fenton-like catalyst. Chinese Journal of Chemical Engineering. 2015;**23**:1698-1704

[49] Wu X, Huang M, Zhou T, Mao J. Recognizing removal of norfloxacin by novel magnetic molecular imprinted chitosan/c-Fe_2O_3 composites: Selective adsorption mechanisms, practical application and regeneration. Separation and Purification Technology. 2016;**165**:92-100

[50] Gutierrez-Climente R, Gomez-Caballero A, Guerreiro A, Garcia-Mutio D, Unceta N, Aránzazu Goicolea M, Barrio RJ. Molecularly imprinted nanoparticles grafted to porous silica as chiral selectors in liquid chromatography. Journal of Chromatography A. 2017;**1508**:53-64

[51] Torkashvand M, Gholivand MB, Malekzadeh G. Construction of a new electrochemical sensor based on molecular imprinting recognition sites on multiwall carbon nanotube surface for analysis of ceftazidime in real samples. Sensors and Actuators B. 2016;**231**:759-767

[52] Prasad BB, Jaiswal S, Singh K. Ultra-trace analysis of d-and l-aspartic acid applying one-by one approach on a dual imprinted electrochemical sensor. Sensors and Actuators B. 2017;**240**:631-639

[53] Liu X, Lv P, Yao G, Ma C, Huo P, Yan Y. Microwave-assisted synthesis of selective degradation photocatalyst by surface molecular imprinting method for the degradation of tetracycline onto Cl-TiO_2. Chemical Engineering Journal. 2013;**217**:398-406

[54] de Escobar CC, Lansarin MA, dos Santos JHZ. Synthesis of molecularly imprinted photocatalysts containing low TiO_2 loading: Evaluation for the degradation of pharmaceuticals. Journal of Hazardous Materials. 2016;**306**:359-366

[55] Meischl F, Schemeth D, Harder M, Kopfle N, Tessadri R, Rainer M. Synthesis and evaluation of a novel molecularly imprinted polymer for the selective isolation of acetylsalicylic acid from aqueous solutions. Journal of Environmental Chemical Engineering. 2016;(**4**):4083-4090

[56] Lei J, Tan T. Enantioselective separation of naproxen and investigation of affinity chromatography model using molecular imprinting. Biochemical Engineering Journal. 2002;**11**:175-179

[57] Azodi-Deilami S, Najafabadi AH, Asadi E, Abdouss M, Kordestani D. Magnetic molecularly imprinted polymer nanoparticles for the solid-phase extraction of paracetamol from plasma samples, followed its determination by HPLC. Microchimica Acta. 2014; **181**:1823-1832

[58] Buchberger WW. Current approaches to trace analysis of pharmaceuticals and personal care products in the environment. Journal of Chromatography A. 2011;**1218**:603-618

[59] Himmelsbach M, Buchberger W, Miesbauer M. Determination of pharmaceutical drug residues on suspended particulate material in surface water. International Journal of Environmental Analytical Chemistry. 2003;**83**:481-486

[60] Farrington K, Regan F. Molecularly imprinted sol gel for ibuprofen: An analytical study of the factors influencing selectivity. Talanta. 2009;**78**:653-659

[61] Madikizela LM, Chimuka L. Occurrence of naproxen, ibuprofen, and diclofenac residues in wastewater and river water of KwaZulu-Natal Province in South Africa. Environmental Monitoring and Assessment. 2017;**189**:348-360

[62] Javanbakht M, Attaran AM, Namjumanesh MH, Esfandyari-Manesh M, Akbari-adergani B. Solid-phase extraction of tramadol from plasma and urine samples using a novel water-compatible molecularly imprinted polymer. Journal of Chromatography B. 2010;**878**:1700-1706

[63] Terzopoulou Z, Papageorgiou M, Kyzas GZ, Bikiaris DN, Lambropoulou DA. Preparation of molecularly imprinted solid-phase microextraction fiber for the selective removal and extraction of the antiviral drug abacavir in environmental and biological matrices. Analytica Chimica Acta. 2016;**913**:63-75

[64] Matongo S, Birungi G, Moodley B, Ndungu P. Pharmaceutical residues in water and sediment of Msunduzi River, KwaZulu-Natal, South Africa. Chemosphere. 2015;**134**:133-140

[65] Pourfarzib M, Shekarchi M, Rastegar H, Akbari-Adergani B, Mehramizi A, Dinarvand R. Molecularly imprinted nanoparticles prepared by miniemulsion polymerization as a sorbent for selective extraction and purification of efavirenz from human serum and urine. Journal of Chromatography B. 2015;**974**:1-8

[66] Chianella I, Karim K, Piletska EV, Preston C, Piletsky SA. Computational design and synthesis of molecularly imprinted polymers with high binding capacity for pharmaceutical applications-model case: Adsorbent for abacavir. Analytica Chimica Acta. 2006;**559**:73-78

[67] Duy SV, Lefebvre-Tournier I, Pichon V, Hugon-Chapuis F, Puy JY, Perigaud C. Molecularly imprinted polymer for analysis of zidovudine and stavudine in human serum by liquid chromatography–mass spectrometry. Journal of Chromatography B. 2009;**877**:1101-1108

[68] Beltran A, Caro E, Marce RM, Cormack PAG, Sherrington DC, Borrull F. Synthesis and application of a carbamazepine-imprinted polymer for solid-phase extraction from urine and wastewater. Analytica Chimica Acta. 2007;**597**:6-11

[69] Kyzas GZ, Nanaki SG, Koltsakidou A, Papageorgiou M, Kechagia M, Bikiaris DN, Lambropoulou DA. Effectively designed molecularly imprinted polymers for selective isolation of the antidiabetic drug metformin and its transformation product guanylurea from aqueous media. Analytica Chimica Acta. 2015;**866**:27-40

[70] Haginaka J, Sanbe H. Uniformly sized molecularly imprinted polymer for (S)-naproxen: Retention and molecular recognition properties in aqueous mobile phase. Journal of Chromatography A. 2001;**913**:141-146

[71] Haginaka J, Sakai Y. Uniform-sized molecularly imprinted polymer material for (S)-propranolol. Journal of Pharmaceutical and Biomedical Analysis. 2000;**22**:899-907

Chapter 4

Plasma Polymerization for Tissue Engineering Purposes

Gaelle Aziz, Rouba Ghobeira, Rino Morent and
Nathalie De Geyter

Additional information is available at the end of the chapter

http://dx.doi.org/10.5772/intechopen.72293

Abstract

The ability of non-equilibrium plasmas to modify surfaces has been known for many years. And a promising way to perform surface modifications without altering the bulk properties is plasma polymerization since this technique is versatile and can be applied to a wide range of materials. Plasma polymer films usually show good biocompatibility when compared to classical biomaterials. The possible biomedical use of plasma polymers motivates the study of their behavior during storage and in aqueous environment. Therefore, it is of major importance to understand the change of properties of these plasma polymers over time and when in contact with certain fluids. Recently, plasma polymer gradients (surfaces that display a change in at least one physicochemical property over distance) have attracted significant attention from the biomedical filed where the interaction of cells with a material surface is of major interest. This chapter discusses biomaterial functionalization via plasma polymerization focusing on their use in the biomedical field as well as their aging and stability behaviors. Plasma polymer gradients as valuable tools to investigate cell-surface interactions will also be reviewed.

Keywords: biomaterial, plasma polymer, surface gradient, stability, aging

1. Introduction

1.1. Tissue engineering

Tissue engineering (TE) was first expressed at the NSF "National Science Foundation" workshop in 1987 by Dr. Fung. TE was later described as the application of engineering and life sciences to better understand the correlations between the structure and the function of tissues as well as the development of replacements for the restoration, preservation and/or enhancement of tissue functions [1].

But it was not until 1993 that Langer and Vacanti gave TE the classical definition of: "an interdisciplinary field that applies the principles of engineering and the life sciences toward the development of biological substitutes that restore, maintain, or improve the tissue function [2]." Various, more or less similar TE definitions can be found in the literature. Moreover, since this is a relatively new field, specific definitions are not always given and may stretch from decellularized matrices to cellular implants.

In tissue engineering, biomaterials must possess appropriate surface properties for better cell-material interactions. In addition, biomaterials should possess appropriate bulk properties to function properly in a bio-environment. Therefore, a suitable approach is to select a biomaterial having good bulk properties and enhance its surface properties using a preferential surface treatment [3, 4]. In this way, one can obtain an "ideal" biomaterial with selective surface properties that are decoupled from its bulk properties and avert the need to develop completely new materials which is quite costly and time-consuming.

In the past few decades, tailoring materials surface properties has been extensively performed using various modification techniques such as chemical treatments and etching, ozone treatment, UV radiation, and plasma treatments [5–10].

Plasma surface treatments are most promising due to the speed and uniformity of modification, their chemical flexibility and positive environmental impact [11, 12]. Various types of plasma surface modification technologies have been used to modify materials by incorporating a variety of functional groups on their surfaces; this is done to improve the surface energy, wettability, adhesion, and bioactive response [13, 14].

1.2. Plasma: a brief introduction and historical background

Plasma is defined as the fourth state of matter in the sequence: solid, liquid, gas, and plasma. The transition between these different respective states can occur by increasing the temperature of the material under consideration.

This state of matter was first described in 1879 by Crookes as "a world where matter may exist in a fourth state." Later, in 1928, this state of ionized gas was eventually given its name "plasma" by Irving Langmuir, when he introduced it in his studies of electrified gases in vacuum tubes [15]. Plasmas can be natural such as lightning, polar light and the stars or man-made. Therefore, without being aware, every person has faced various forms of plasma. Man-made plasma can be generated in laboratories by combustion, flames, lasers or controlled nuclear reactions. But, in the field of plasma polymerization, most plasma are generated and sustained using an electrical discharge.

Plasma is generally formed when gas atoms are subjected to a high enough thermal or electrical energy. Subjected to energy, gas atoms become ions by releasing some of their electrons. Radicals are then created by electron-molecule collisions and bond breaks in molecules. Some excited species will also be created by energy adsorption which will generate photons. This unique mixture of electrons, ions, radicals, photons and neutrals constitute the so-called plasma [16, 17].

Plasmas are classified as thermal "equilibrium" and non-thermal "non-equilibrium" based on the relative temperatures of electrons, ions and neutrals.

In a non-thermal or cold plasma, the electron temperature (≈ 10,000°C) is much higher than the gas temperature (< 200°C), whereas, in a thermal or hot plasma, the electron temperature is very close to that of the heavy particles.

Plasmas used in the field of plasma polymerization are usually cold plasmas since it involves heat-sensitive materials [18, 19].

2. Plasma modification of surfaces

The ability of non-thermal plasmas to dramatically modify surfaces properties has been known for over 25 years. Plasma treatments allow surface modification of polymeric materials without altering their bulk properties [20–22]. These plasma processes can be categorized into 3 major types of reaction: plasma activation, post-irradiation grafting (briefly discussed in the next sections), and plasma polymerization (the focus of this chapter).

2.1. Plasma activation

In plasma activation, surface modification is done by exposure to non-polymer forming plasmas. The active species in the plasma can bombard the polymeric surface and break covalent bonds thus leading to radical formation. These radicals can subsequently react with other species in the plasma to form functional groups. In this way depending on the selected plasma gas, different functional groups such as carbonyl, carboxylic acid, hydroxyl, and amine functional groups can be added on the surface thus making it more hydrophilic [23–25]. It is believed that radical species rather than ions or electrons are most important in this type of modification [26].

2.2. Plasma polymerization

Observations of organic compounds formed in a hydrocarbon based plasma discharge dates back to 1874 [27]. These deposits were considered to be undesirable by-products. However, in the 1960s [28–30], studies of plasma polymerization started and were completed by considerable advances in polymer science [31]. Plasma polymerization was defined as "the formation of polymeric materials under the influence of plasma" [31]. Nevertheless, the real potential of plasma polymerization was not uncovered until only the past two decades. Nowadays, plasma polymerization is known as a very valuable surface modification technique.

During the process of plasma polymerization, high energy electrons as well as UV will ionize the precursor molecules [32–34]; this leads to radicals which are highly unstable and reactive species that will interact and bond with one another and deposit on the substrate thus forming a coating on its surface. Plasma will also lead to bond breaks on the substrate surface thus creating radicals. These will interact with the precursor's radicals acting as anchor sites which enhances the plasma polymerized coating stability.

During plasma polymerization, two processes occur simultaneously: - ablation (removal of surface molecules) and – polymerization (surface monomer deposition). These two processes are in competition and their interaction and co-existence in plasma is well known [35].

Plasma polymerization is very complex and versatile since various parameters such as discharge power, treatment time, precursor type and concentration can affect the physic-chemical characteristics of the deposited coating. Moreover, different reactive species can be formed in the plasma depending on the used dilution gas (e.g., helium, argon, air or nitrogen), which also affects the characteristics of the coating [25, 36–38].

Advantages of plasma polymerization include the following:

1) Ultra-thin film formation

2) Good adhesion to the substrate material and deposition is independent on the structure or type of the substrate

3) Relatively good chemical stability and physical durability of the coatings

4) Various precursors can be chosen which leads to a vast array of surface functionalization (monomers used do not have to contain a double bond for the polymerization to proceed)

5) Many process parameters can be used thus providing great diversity of surface modifications

6) The obtained coatings are more or less uniform

Nevertheless, plasma polymerization also presents several disadvantages:

1) System dependency

2) Scaling up and converting it into a continuous process could present some technical challenges

3) The specific roles of each plasma component are difficult to separate and analyze

4) It is hard to predict the exact surface characteristics of the deposited plasma polymer especially when complex molecules are used

5) Coating multi-functionality can also be an issue

6) Everything in the coating range of the plasma can become part of the coating

However, despite its disadvantages and focusing on its numerous advantages, plasma polymerization has rapidly developed during the past decades and is now used for various applications.

Plasma polymers

Plasma polymers are markedly different from conventional polymers. Conventional polymers have a well-defined structure of repeating units that corresponds to the used monomer. Whereas, plasma polymers are crosslinked, randomly structured deposits obtained from the fragmentation and recombination of monomers within an electric discharge.

During plasma polymerization, the active species fragments the organic precursor (monomer), thus creating radicals that can recombine both in the plasma and on the substrate surface forming a crosslinked so-called plasma polymer coating/film on the substrate surface.

As to the film chemical structure, during this process, partial loss of functional groups occurs (fragmentation) in a system/process dependent way. Moreover, not all radicals will react and some will be trapped in the plasma polymer network [41]. As a consequence, the elemental composition of plasma polymers differs from that of conventional polymers prepared from the same monomer. For example, the elemental composition of conventionally polymerized polyethylene $(C_2H_4)_n$ is equal to that of the monomer (C_2H_4); however, in plasma polymerized ethylene the hydrogen concentration is lower (radical formation by -H bond scission) and oxygen is incorporated in the plasma polymer (by reaction with the formed radicals).

Hence, the material obtained from plasma polymerization is very different than that obtained by conventional polymerization of the same monomer [39].

2.3. Post-irradiation grafting

Surface modification via polymer coatings is also frequently done by surface grafting methods which are often referred to as "plasma-induced graft (co)polymerization." This is a two-step process. In the first step, the surface is exposed to air plasma or subjected to an ozone treatment which creates peroxide groups at the surface. Other non-oxygen containing plasma can also be used (e.g., Ar or He plasma) followed by atmosphere or O_2 exposition; created radicals will then form peroxides and hydroperoxides. In the second step, the formed functionalities are used to initiate a polymerization reaction by contact with the monomer molecules. Each functionality being a potential initiating site, the number of created (hydro)peroxides has a significant effect on the surface grafting density.

3. Cellular response to surfaces

For the cells, the surface is the most important part of the material. Cell-biomaterial interactions depend on the surface energy, chemical composition and surface morphology [40, 41]. Moreover, cell growth, spreading and viability were shown to be closely linked to their adhesion on the surface [42]. Consequently, suitable surface properties contribute to better cell adhesion and subsequent proliferation. It is well established that for numerous cell types surface wettability is a paramount factor that influences cell adhesion, with this being more favorable on hydrophilic surfaces compared to hydrophobic surfaces. **Figure 1** shows micrographs of fibroblast cells cultured on untreated (hydrophobic) and argon plasma treated (hydrophilic) UHMWPE substrate. As seen in this **Figure 1**, cells are significantly more spread on the hydrophilic treated surface compared to the hydrophobic untreated one [43]. Additionally, surface charge has also been shown to have a significant influence on cell adhesion [44, 45].

A promising way to achieve optimal surface attributes (e.g., surface wettability and charge) is plasma polymerization which has already been used successfully to enhance cell adhesion on various substrates.

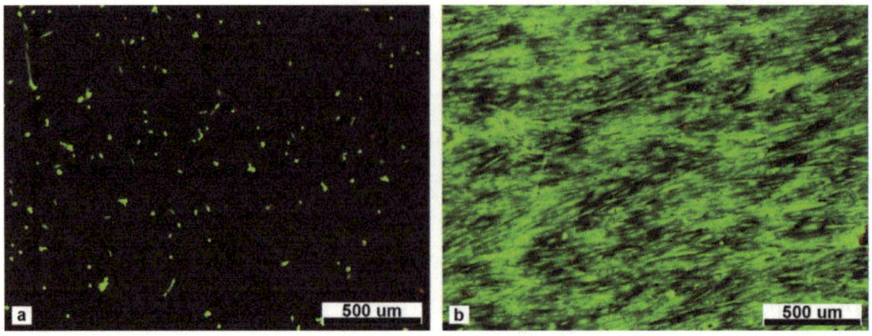

Figure 1. Overall cell morphology of fibroblasts cultured on (a) untreated and (b) argon plasma-treated UHMWPE.

In this section, previous works on the bio-application of plasma polymers and their interactions with cells will be reviewed.

Biological applications of plasma polymers

Plasma polymerization is a convenient way to introduce desired functional groups on a surface. Plasma polymers are frequently used to immobilize biomolecules and enhance cell adhesion. NH_2 and COOH based plasma polymers are most commonly used since these groups are known for their good chemical reactivity. Moreover, in aqueous solution at physiological pH value, amino/carboxyl groups can introduce a positive/negative charge to the surface thus increasing its affinity for biological components [46–48]. For example, DNA [49, 50], heparin [46], glucose oxidase [51], and collagen [52] have been immobilized on amine or carboxyl based plasma polymers. Hydroxyl and aldehyde groups have also been used to bind heparin [53] and collagen/albumin [54], respectively. However, plasma polymers with these groups are less extensively investigated due to their lower reactivity.

For the effect of plasma polymers on cell attachment and proliferation, various studies on different substrates using numerous plasma media and cell types have been performed. A summary of some of these studies is presented in **Table 1**.

Furthermore, plasma polymer films were used for bacterial adhesion and biofilm prevention by coating the surface with a suitable antibacterial agent (e.g., silver nanoparticle).

Xiaolong et al. [55] produced PET fabrics with antibacterial properties by depositing a plasma polymer organosilicon film where silver nanoparticles were incorporated. A similar study was also conducted on PET meshes by plasma polymerization of acrylic acid followed by incorporation of Ag nanoparticles [56]. Results showed excellent mesh antibacterial properties with a decrease of more than 99.7% in bacterial concentration compared to an untreated mesh. In another study, Degoutin et al. [57] used plasma to graft acrylic acid onto nonwoven polypropylene and the carboxyl groups were used to immobilize an antibiotic "gentamicin." Results showed a 99% efficacy against *E. coli* bacteria.

These results and discussions strongly support the idea that polymer coatings represent a very promising way to modify a biomaterial surface in order to adapt it to a specific biomedical application.

http://dx.doi.org/10.5772/intechopen.72293

Substrates	Plasma media	Cell lines	Observations	Refs
Si, PS, PET	Acrylic acid	Fibroblast	Enhanced fibroblast adhesion	[48]
PET, TCPS	Acrylic acid	3 T3 murine	Improved cell adhesion	[58]
PET	Acrylic acid	Smooth muscle cells	Immobilization of proteins and cell growth	[52]
Glass coverslips, PLGA	Acrylic acid	Caco-2	Mammalian cell sheet formation	[59]
PET	C_2F_4	3 T3 fibroblast	Cell adhesion, growth and proliferation	[60]
PET	C_2F_4	NCTC 2544, 3 T3 fibroblast and MG-63	Good cell adhesion and proliferation	[61]
Glass	Acrylic acid	Leukemia cells	Lower cell growth (60% reduction)	[62]
PCL, PLA	Allylamine, C_2H_4/N_2	Osteoblast, 3 T3 fibroblast	Increased cell metabolic activity and improved cell colonization in the core region of the scaffold	[63]
Titanium alloy	Allylamine and ethylene diamine	MG63	Improved cell adhesion, function and spreading	[64]
LUX tissue culture dishes	Acrylic acid	Rat osteosacroma cells	Improved cell adhesion	[65]
PET	Acrylic acid and allylamine	Human neuroblastoma cells	Improved cell adhesion, differentiation and maturation	[66]
PEEK	Acrylic acid	MC3T3-E1	Improved cell adhesion, spreading and proliferation	[67]
PET, Si	C_2F_4	CVEC	Enhanced endothelial cell response; increased cell attachment, spreading and viability	[68]
PCL, PLLA	O_2, acrylic acid, allylamine	MC3T3-E1	Improved protein adsorption and cell attachment	[69]
Ti	Allylamine	MG63	Improved adhesion and cell functions	[45]
PET	Allylamine	Human skin fibroblast	Improved cell attachment, viability and metabolic activity	[70]
PS	Isopropyl alcohol	Fibroblast	Enhanced cell attachment and proliferation	[71]

Table 1. Summary of some of the studies on plasma surface modification of materials and their effect on cell adhesion and growth.

4. Effect of aqueous environments on plasma polymers

For biomedical applications, the effect of water on the plasma polymer films is of particular importance.

Immersed in a solvent, plasma polymers can be subject to numerous processes such as:

- delamination from the substrate
- detachment of oligomers
- swelling
- reaction with the solvent

However, not many studies focus on the physic-chemical changes that happen to the plasma polymer films after exposure to aqueous environments.

Plasma polymer stability behavior depends on the type of the polymer. Muir et al. [72] studied the penetration of water into the films and characterized the swelling of allylamine (Aam) and heptylamine (HA) plasma polymers. When immersed in water, the plasma polymerized Aam film (ppAam) was found to swell by 5% and to contain 3% water whereas the ppHA film did not appear to swell but contained 5% water. The swelling characteristics of other plasma polymers have also been reported [73–75].

Moreover, the degree of swelling strongly depends on the plasma process parameters. Zhang et al. [73], demonstrated that ppAam deposited at 20 W only shows a small degree of swelling while ppAam deposited at 5 W shows a large degree of swelling. This is due to the fact that at low powers the plasma polymer contains a large number of oligomers which are not covalently bound to the film; these oligomers can thus be readily extracted in the solvent. In fact, when studying the morphology of ppHA, Vasilev et al. [76] found that pores of several nanometers in diameter were formed after ppHA has been immersed in water for 24 h (see **Figure 2**). And the dimension of the pores was found to depend on the deposition conditions with larger pores obtained at lower powers (see **Figure 3**). This was attributed to oligomer water extraction after low molecular weight fragments were detected in the water. This results in the formation of gaps in the film and leads to ruptures of the polymer chains thus forming the observed porosity.

Förch et al. [77] found that for ppAam, the roughness of the polymer film increased from 0.85 to 1.26 nm after soaking in water which was attributed to the swelling of the film in water;

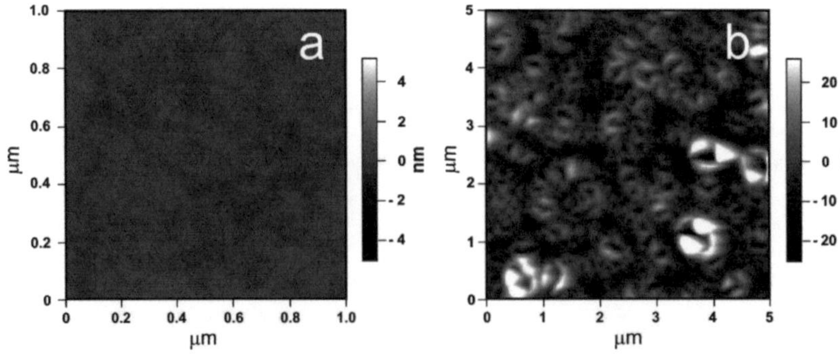

Figure 2. AFM topographic images of HA plasma polymer films deposited with a power of 20 W: (a) as deposited, (b) after immersion in water for 24 h.

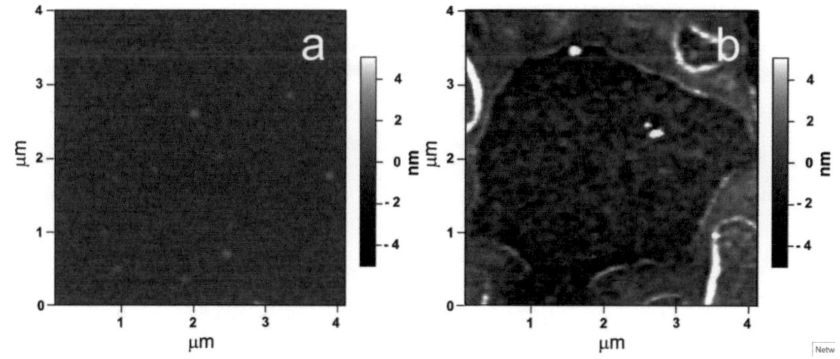

Figure 3. AFM images of HA plasma polymer films deposited with a power of 50 W: (a) as deposited or (b) after immersion in water for 24 h.

whereas, Tarasova et al. [78] used XPS to study the changes in surface chemistry of ppAam and ppHA after immersion in water for up to 24 h. Results were similar to the ones obtained after these plasma polymers were stored in air, both undergoing rapid oxidation; amine and imine groups were converted to amides with an increase of C—O and C=O groups.

In order to improve plasma polymers stability, studies on the interaction of plasma polymer with the aqueous environment as a function of plasma deposition parameters have been conducted. Optimizing these parameters was shown to be very important in reducing the induced changes [77, 79]. Moreover, substrate pretreatment for cleaning or activation was also shown to prevent the delamination of the polymer film [77].

However, enhancement of plasma polymer stability is still insufficiently studied and more effort is still needed for a precise stability evaluation and quantification.

5. Surface aging

It is widely known that the enhancement in surface wettability obtained after plasma activation processes changes with storage time. This phenomenon is referred to as aging or hydrophobic recovery and is due to the tendency of a surface to minimize its surface energy by reverting to its original structure. This leads to a loss of surface polar functional groups that re-orientate to the bulk [18]. Therefore, in the case of plasma activation, in order to avoid the adverse effect of aging, it is advisable to only use freshly prepared samples.

On the other hand, plasma surface grafted polymers and plasma polymerized films show much less modifications after storage in ambient air and are thus considered comparatively stable in time. However, research on plasma polymers show that they are susceptible to oxidation upon storage in air [31]. Since these coatings have shown great potential for many applications including biomedical ones, several studies have been done to better understand this so-called aging process and therefore further evaluate the relevancy of plasma polymers. And since most products are usually stored for a certain period before they are used, the film properties at the time of use are usually considered more important than immediately after treatment.

Major advancements in the understanding of oxidative reaction mechanisms that occur during plasma polymer aging have been made by Gengenbach et al. [80–83].

This was done using XPS, FTIR spectroscopy and contact angle goniometry characterization techniques which allowed significant perception of the eventual surface compositional changes.

Their studies included detailed oxidation investigations of hydrocarbon based plasma polymers [80], fluorocarbon coatings [81], nitrogen coatings [82] and other plasma deposited films [83]. Results showed that the aging process was due to the reaction of ambient oxygen with the residual radicals present in plasma polymers; ESR spectroscopy showed that the free radicals detected in freshly deposited plasma films slowly disappear upon storage in air. Results also showed that, the kinetics, mechanisms and formed oxidative products during aging depend on many factors, such as the structure of the film, the type of functional groups and the mobility of the surface.

6. Plasma polymer gradients

After accomplishing significant advancements in the biofunctionalization of surfaces by different chemical and physical homogeneous modifications, a growing research interest is being shifted toward the development of gradient surfaces presenting graded wettability, chemistry, biomolecule density and nanoparticle distribution [84]. This interest stems from the fact that many essential and poorly understood biological activities are driven by such gradients. For instance, chemotaxis mediate a number of physiological processes such as leukocyte recruitment to the infection site, guiding of neuronal and glial cells during nervous system development or regeneration and cancer metastasis. Moreover well-ordered gradient distribution of specific functional groups, extracellular matrix components, signaling biomolecules and even topographical cues induce particular cell type proliferation, migration and differentiation [85–87]. Besides their biological importance, gradients are also powerful for high throughput screening in several applications such as biomaterial development, tissue engineering and sensors, in the sense that a single sample designed with a gradient surface is used to procure multiple data points. This reduces dramatically the number of samples and cells, eliminates inaccuracies triggered by sample reproducibility and speeds up the analysis [84, 88, 89].

Different approaches are commonly adopted to create surface gradients such as self-assembled monolayers (SAMs), grafting on hydrogels and Boyden chambers and filters. Several limitations are associated with these traditional methods including the substrate dependency (e.g., gold-coated surface is required for SAMs), the short term gradient "shelf-life," the restricted chemistries that can be obtained and the long experimental timing [46, 90, 91]. As an alternative, applying high energy source plasmas that are associated with many advantages such as the absence of solvent, the specificity and the substrate independency has shown great successes in the generation of gradient surfaces. In 1989, Witt el al. were one of the first groups if not the first, to generate wettability gradients on polyethylene, polystyrene, polydimethylsiloxane, and polytetrafluoroethylene by a radio frequency (RF) plasma activation operating in oxygen, ammonia and sulfur dioxide atmospheres. A special gradient apparatus consisting of an aluminum box with a translating cover and two aluminum plates serving as electrodes, was designed for this purpose. During the treatment, the cover is retracted with a constant velocity automatically controlled by a microprocessor driving the stepping motor. This movement linearly increases the plasma exposure time over the sample length. As a result, water contact angles (WCA) increased along the length of the sample thus ensuring the presence of a wettability gradient. Moreover, a wide range of wettability gradients is obtained by varying the gas, the radio frequency power, the cover retraction velocity and the plasma exposure time. This study highlighted the high flexibility of the plasma treatment to generate gradients with defined length and magnitude and pointed out, by using several substrates, that the process is substrate independent [92]. Therefore, a steep rise in literature focusing on the generation of gradients by plasma activation followed. However, the different plasma activation methods that were described are mainly limited to the production of wettability gradients with a relatively restricted control over the chemical group specific incorporation. Other concerns include the aging effect of the treated surfaces due to the reorientation of the incorporated groups away from the surface when the environment is thermodynamically unfavorable and the roughening of the surface due to the plasma etching effect [92, 93]. Consequently, the interest was shifted toward the generation of polymer gradients via plasma polymerization to be able to control more precisely the functional group nature and densities, the gradient stability and the gradient shape. Nevertheless, it was until 2003 that the first method enabling the deposition of controllable horizontal plasma chemical gradients was described by Whittle et al. [93] and was subsequently adopted as it is or with some amendments by several other groups [91, 94]. In their study, Whittle et al. created hydrocarbon/carboxyl and amine/carboxyl functionality gradients on glass substrates over a distance of 11 mm. Instead of using the traditional cylindrical plasma reactor, a RF glow discharge T-shaped reactor presenting a drawer as sample holder was used [93, 95]. As a first step, an amine coating was deposited on the whole glass substrate by performing a plasma polymerization using allylamine (Aam) monomers as precursors and a continuous power of 10 W while the drawer was fully extended. An underlayer presenting a good adhesion was thus formed for the subsequent gradient deposition. In the second step, the power was decreased to 5 W and a plasma polymerization was performed while the drawer was slowly closed at a constant velocity of 1 mm/min along with a controlled change in the plasma composition over time. This was performed by introducing acrylic acid (Aac) as the second monomer while decreasing instantaneously the flow rate of Aam by 4 cm^3 stp/min. For the deposition of hydrocarbon/carboxyl gradients, the same procedure was followed but with the use of octa-1,7-diene instead of Aam precursors. The obtained plasma polymerized surfaces were characterized by X-ray photoelectron spectroscopy (XPS) and chemical derivatization of acid functionalities using trifluoroethanol. A gradual increase in the concentration of acid functionalities was observed in the case of hydrocarbon/carboxyl gradients and an increase of acid and amine functionalities was attained in opposite directions in the case of the amine/carboxyl gradients. These findings demonstrated the power of this first-hand methodology to successfully generate plasma polymer gradients that can subsequently allow the grafting of a broad range of biochemical entities in a spatially structured manner [93]. Surface engineers waited around 3 years after the study of Whittle to begin their investigations regarding the grafting of biomolecules and the cell-biomaterial interactions when a plasma polymer gradient is implicated. Moreover, several other methods generating plasma polymer gradients were described with a distinctive focus on amine and carboxylic acid being the most 2 extensively considered functionalities in the subsequent literature of gradient plasma polymerization. In what follows, an overview on the achievements of these carboxylic acid and amine plasma gradients in several tissue engineering and biomedical applications will be given.

6.1. Surface plasma polymer gradient of carboxylic acid functionalities

In 2006, Parry et al. [91] performed a plasma copolymerization of Aac and octadiene (OD) based on the mechanism described by Whittle et al. [93] but with a modification of the setup in a way allowing the production of 20 similar gradients at a time. Up to 20 substrates could thus be placed in the redesigned RF plasma reactor and moved under a slot by an automated stepper motor in 250 μm paces at a rate of 750 μm/min. Simultaneously, a controlled composition of the monomer mixture is sent to the chamber via two computer-regulated valves. A thorough characterization of the surface gradient was executed by angle resolved x-ray photoelectron spectroscopy (ARXPS) that showed in great details how acid functionalities changed on different positions of the gradient and highlighted the presence of vertical changes especially when it comes to the plasma polymer thickness. An assay investigating the passive adsorption of immunoglobulin G (IgG) as a function of the acid surface density was supplemented to the study to be, to the best of our knowledge, the first reported biological assay done on plasma polymer gradients. ARXPS measurements showed that IgG was by far more absorbed on the OD gradient end and that IgG amount decreased gradually as the concentration of Aac increased thus creating an IgG gradient [91]. In 2009, Walker et al. [96] also deposited a gradient of OD/Aac on coverslips using the plasma deposition/masking method of Whittle but this time with a renovated protocol permitting the generation of submillimeter-scale gradients instead of millimeter scale length. In the updated method, OD was constantly fed to the reactor as the slot moves across the substrate surface, then it was brusquely turned off and a pulse of Aac was launched. The scale length and density of the carboxylic groups were thus tailored by varying the pulse width of Aac. The obtained gradient surface was used to immobilize the intercellular signaling molecule delta-like-1 Dll 1, a factor enhancing stem cells self-renewal and preventing cell differentiation which is an issue to be considered when developing cell therapy technologies. Since tiny changes in surface properties can considerably affect the stem cell behavior either by enhancing the commitment path toward their differentiation to particular cell types or by maintaining and stabilizing the stem cell pluripotent phenotype, concentration-based factor and chemical group gradients are highly expedient to study stem cells. Instead of directly grafting Dll 1 factor on the generated gradient, a mouse monoclonal antimyc-tag (9E10) antibody is covalently coupled, then Dll 1 is immobilized on the antibodies thus avoiding the alteration of its biological activity by separating it from the solid surface. A visualization of the Dll 1 gradients was performed by binding a rabbit anti-Dll-1 antibody and then introducing a colloidal gold-conjugated secondary antibodies. Several Dll 1 gradients with different slopes and end points were obtained depending on the plasma Aac pulse width adopted during the plasma polymerization (**Figure 4**). During the same year, the first cell tests on plasma gradients were performed by Wells et al. using mouse embryonic stem (ESC) cell lines E14 and R1 in order to examine their pluripotency [97]. OD/Aac gradients were deposited on coverslips using the same setup described by Parry et al. [91]. The degree of cell spreading was studied in function of COOH concentration. Alkaline

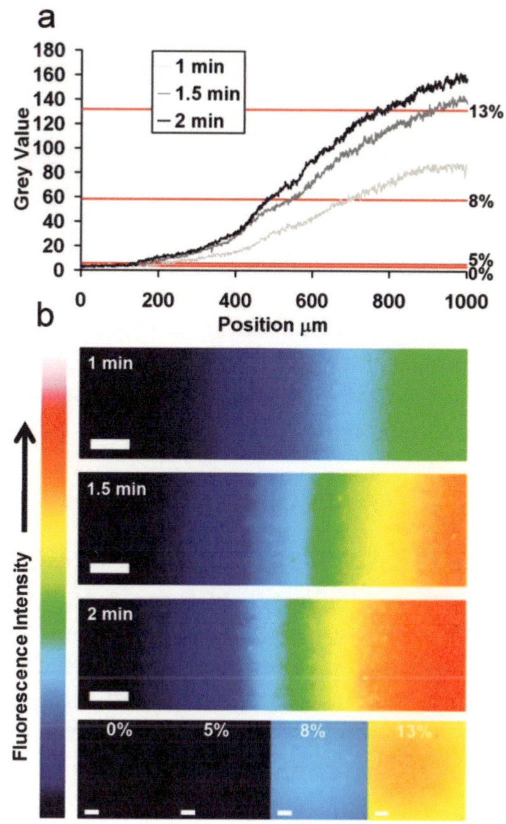

Figure 4. (a) Densitometry results of 9E10 antibodies immobilized on the gradient surface (3 different Aac pulse durations) and visualized by FITC-conjugated secondary antibodies. Horizontal lines show the results of homogeneous surface treatments (b) false color heat maps of the 9E10 antibody gradients. Homogeneous surfaces are presented for comparison. Scale bars: 100 μm.

phosphatase staining showed that cell capacity of self-renewal is preserved when the cell spreading is still below 120 μm^2 [97]. In 2012, in an attempt aiming to make a sweeping statement about this result, Harding et al. [98] used polyethylene oxide (PEO) that is well-known in the biomaterials field to limit protein adsorption and thus cell adhesion and spreading, together with Aac to produce two counter gradients. A RF apparatus consisting of a cylindrical glass chamber was used for the plasma copolymerization. As a first step, an OD layer then an Aac layer were deposited on the substrate since a unique Aac deposition resulted in the coating dissolution in water. Then a mask 12 °titled in respect to the surface was employed to deposit a PEO-like gradient by using the monomer diethylene glycol dimethyl ether (DG) as a precursor. A successful fabrication of AA-DG plasma polymer gradient was revealed by XPS, profilometry and infrared microscopy mapping. The gradient could be easily altered by changing the plasma process parameters. Mouse ESC were cultured on the gradient surfaces, then immunocytochemical stainings of the stem cell markers Oct4 and alkaline phosphatase were performed. Results showed a low cell adhesion and colony formation on the DG rich end and an increased colony size and decreased stem cell marker expression on the COOH rich end, thus supporting the hypothesis stating that cellular spreading influences the fate toward cell differentiation or self-renewal. The same method using a tilted mask was then applied by Wang et al. [99] in 2014 to create the same Aac-DG gradients but also Aac-OD gradients by firstly depositing OD uniformly then using the tilted mask to deposit Aac. Attachment and differentiation of rat bone marrow mesenchymal stem cells (rBMSCs) into adipogenic and osteogenic lineages were investigated on both gradients. After 24 h of cell culture, a gradient in cell density was observed on the substrate with a decreased cell adhesion on DG and OD rich ends. The obtained cell density gradient vanished on Aac-OD gradient after 6 days but not on Aac-Dg gradient, thus suggesting the long-term efficacy of the later gradient. Cell colonies containing bone nodules were detected on this gradient especially on the Aac rich ends but not on the DG rich end. Moreover, proteins and calcium were not secreted on the DG end implying that osteogenic differentiation is influenced by local cell densities. However, the induction of the cells toward an adipogenic lineage showed that this differentiation is cell density insensitive.

6.2. Surface plasma polymer gradient of amine functionalities

In addition to COOH functionalities, NH_2 groups were also shown to be very powerful in influencing a wide range of particular cell type performances such as adhesion, proliferation, migration and differentiation. Therefore, when the research community started investigating surface gradients, a distinctive focus was directed toward the production of amine gradients and their use in several biomaterial and tissue engineering applications [100]. To the best of our knowledge, all COOH plasma polymer gradients described so far were only deposited on flat substrates, however some amine plasma polymer gradients were deposited on 3D scaffolds. For instance, in 2006 Barry et al. [101] thought of generating an amine gradient on poly(D,L-lactic acid) 3D porous scaffolds in order to solve the common problem of the highly disproportionate cell colonization on the scaffold periphery in comparison to the hardly accessible scaffold center that remains poorly colonized and supplied by nutrients. This issue was solved by plasma polymerizing hexane, known to be resistant to cellular adhesion, on the periphery of the scaffold while generating an amine plasma polymer coating on the central surface. To do so, a first plasma polymerization step using Aam monomers as precursors was performed, then a second polymerization using the cell-repellent hexane was achieved at lower deposition rate. XPS measurements throughout the whole scaffold showed that when the second hexane polymerization step is absent, a decrease in amine functionalities is observed toward the center. However, when hexane polymerization is introduced, the nitrogen concentration is reduced by 1 to 2% in the periphery thus creating a reversed gradient. After seeding 3 T3 fibroblasts on the treated scaffolds, X-ray micro-computed tomography and scanning electron microscopy revealed a uniform cell distribution throughout the whole scaffold with well spread cells in the center associated with a high production of extracellular matrix (ECM) components. The use of hexane and Aam to create amine gradients was also considered by Zelzer et al. [102] in the subsequent year, but this time on flat glass coverslips. The idea behind the study was to compare between mammalian cell interactions on gradient and on uniformly treated surfaces. A T-shaped borosilicate RF reactor was used to plasma polymerize uniformly an amine coating on glass coverslip using Aam as precursors. Afterward, a poly-hexane was deposited on the poly-Aam coated surface after placing a mask either directly or making use of a spacer clamping the mask at a distance of 0.04 mm from the surface. The direct positioning of the mask resulted in steep gradients while the use of a spacer gave more shallow gradients. Wettability gradients were detected by WCA measurements showing a gradual decrease from 93° to 66°, thus correlating with the gradual increase of N/C ratio. NIH 3 T3 fibroblasts cultured on the gradients surfaces were preferentially adhered and proliferated on the N-rich end with a gradual cell density decrease toward the poly-hexane rich end. Surprisingly, experiments performed on uniform surfaces revealed significant differences in cellular behavior compared to the gradient surfaces, leaving question marks on the use of gradients for high throughput screening. The cell signaling and the protein synthesis might be different between gradient and uniform surfaces since the cell neighboring environment differs. Several subsequent studies involving amine plasma polymer gradients and their general results are summarized in **Table 2**.

Authors (year)	**Plasma reactor/ monomers used**	**Gradient formation method**	**Surface chemical properties**	**Biological assay/ cell type**	**Bioresponsive properties**
Robinson et al. [46]	Cylindrical RF reactor/ Aam- OD	Moving slot with a simultaneous change in the monomer mixture composition	Gradual increase in N/C ratio over a distance of 14 mm	Adsorption of heparin that mimics the heparan sulfate proteoglycans found in all tissue types	-Gradual increase in heparin adsorption parallel to the increase in N/C ratio. -Heparin functionality not correlated with the continuous increase in heparin adsorption
Robinson et al. [103]	T-shaped RF reactor/ Aam- OD	Moving slot with a simultaneous change in the monomer mixture composition	Gradual increase in N/C ratio on washed and unwashed samples highlighting the stability of plasma polymer gradient surfaces	—	—
Harding et al. [87]	T-shaped RF reactor/ Aam- OD	Moving slot with a simultaneous change in the monomer mixture composition	Gradual increase in N/C ratio over a distance of 12 mm	D3 murine embryonic stem cell line culture	-Maximum cell adhesion on the N-rich end -Inverse increase in stem cell marker expression toward the lower N/C ratio. -Correlation between the presence of stem cell markers and the formation of more multilayered and compact cell colonies.

Authors (year)	Plasma reactor/ monomers used	Gradient formation method	Surface chemical properties	Biological assay/ cell type	Bioresponsive properties
Mangindaan et al. [100]	RF reactor/ Aam	Use of a mask with a 1 mm gap on a polypropylene substrate	-Wettability gradient with WCA varying from 15° to 90°. - Gradient over 1 cm of nitrogen content from 5.8 to 16.0% and amine content from 1.98 to 4.03 per 100 carbons.	L-929 fibroblast culture	Continuous increase in the cellular density with more than 2-fold density on N-rich end
Delalat et al. [104]	RF reactor/ Aam-OD	Moving slot with a simultaneous change in the monomer mixture composition	Gradient over 12 mm of nitrogen content from 0 to 12.0%	Mouse embryoid body cell culture	-Highest cell adhesion on the gradient central regions -Increased cell proliferation toward the Aam end. -Cell differentiation toward mesodermic and ectodermic lineages on high nitrogen content regions -No correlation between amine content and endodermal differentiation
Liu et al. [105]	RF reactor/ Aam-OD	Moving slot with a simultaneous change in the monomer mixture composition	-Wettability gradient with WCAs varying from 90° to 70° -Gradual increase in N/C ratio over a distance of 12 mm - Unchanged surface topography	-Adsorption of fluorescein isothiocyanate-labeled bovine serum albumin (BSA) and rhodamine-labeled fibronectin (FN) -Human adipose-derived stem cell culture	-Gradual decrease in the amount of adsorbed BSA from OD toward Aam sides. -Gradual increase in the amount of adsorbed FN from the OD toward the Aam sides. -Increased cell adhesion and spreading toward the Aam side -No difference in cell performances in the absence of serum -Increased osteogenic cell differentiation toward the Aam side -Decreased adipogenic differentiation toward the Aam side.

Table 2. Overview of literature on amine gradient obtained by plasma polymerization and not discussed in the text.

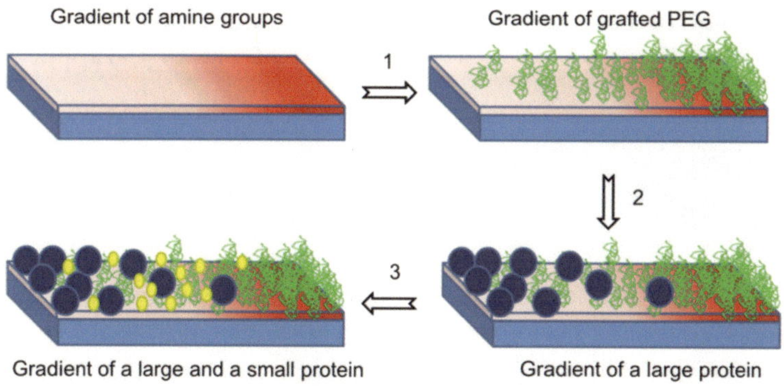

Figure 5. Schematic representation of the creation of two-protein gradient. Step 1. PEG grafting on the amine plasma polymer gradient to generate a PEG density gradient. Step 2. Large proteins adsorption. Step 3. Small protein adsorption.

Since the biological systems in vivo are much more complex than in vitro assays, some authors considered a closer mimicking of the real systems by designing, instead of one dimensional or single protein gradients, 2 protein and 2 dimensional gradients. For instance, in 2009 Vasilev et al. [94] created an Aam-OD gradients on SPRchips or on silicon wafers based on the method described by Whittle et al. [93]. Afterwards, polyethylene glycol (PG), known to be resistant to protein adsorption, was grafted on the amine gradient thus generating a PEG density gradient. The obtained density gradient was then benefited to control the deposition of 2 proteins, namely the large protein fibrinogen and the small protein lysozyme, by differential passive adsorption. A first incubation with the larger protein led to its adsorption on low PEG density regions, then a second incubation with the small lysozyme led to its adsorption only where there is still a "room" for it to adsorb since the previous fibrinogen adsorption passivated gradually the surface. As a result, 2 reversed gradients of 2 proteins could be designed and the method could be generalized to other pairs of small and large proteins (**Figure 5**). In 2013, Mangindaan et al. [90] designed a 2 dimensional amine gradient by performing firstly a plasma polymerization of Aam on a propylene membrane while a mask is placed on top with a gap distance of 1 mm. Subsequently, the same procedure is repeated but after rotating the sample by 90°. WCA measurements showed that both gradients were well controlled by varying the plasma treatment exposure time in each step. L-929 fibroblasts seeded on the treated surfaces adhered and grew proportionally with the amine content on the 2 dimensional gradient with a predominant effect of the gradient created during the initial plasma deposition.

7. Conclusions

From the work presented in this chapter it is clear that plasma polymer coatings are very useful tools for biomaterial surface modification. However, despite the numerous advantages of these coatings for biomaterial advancements, their aging and stability remain an issue that requires further investigations and considerations. More attention and focus on these aspects can make plasma polymerization become one of the most used and important surface modification techniques. Plasma polymer gradients are also very promising for biological applications and many advances in the area of plasma uses can be made by developing such coatings.

Acknowledgements

This research has received funding from the European Research Council (ERC) under the European Union's Seventh Framework Program (FP/2007-2013)/ERC Grant Agreement n. 335929 (PLASMATS).

Author details

Gaelle Aziz*, Rouba Ghobeira, Rino Morent and Nathalie De Geyter

*Address all correspondence to: gaelle.aziz@ugent.be

Department of Applied Physics, Research Unit Plasma Technology (RUPT), Faculty of Engineering & Architecture, Ghent University, Ghent, Belgium

References

[1] Nair LS, Bhattacharyya S, Laurencin CT. Nanotechnology and tissue engineering: The scaffold based approach Nanotechnologies for the Life Sciences. 2006;**1**:4-23

[2] Langer R, Vacanti JP. Tissue engineering. Science. 1993;**260**(5110):920-926

[3] Gupta AK, Gupta M. Synthesis and surface engineering of iron oxide nanoparticles for biomedical applications. Biomaterials. 2005;**26**(18):3995-4021

[4] Jayagopal A, Stone GP, Haselton FR. Light-guided surface engineering for biomedical applications. Bioconjugate Chemistry. 2008;**19**(3):792-796

[5] Abenojar J, Torregrosa-Coque R, Martínez MA, Martín-Martínez JM. Surface modifications of polycarbonate (PC) and acrylonitrile butadiene styrene (ABS) copolymer by treatment with atmospheric plasma. Surface and Coatings Technology. 2009;**203**(16):2173-2180

[6] Kaczmarek H, Kowalonek J, Szalla A, Sionkowska A. Surface modification of thin polymeric films by air-plasma or UV-irradiation. Surface Science. 2002;**507**:883-888

[7] Arenholz E, Svorcik V, Kefer T, Heitz J, Bäuerle D. Structure formation in UV-laser ablated poly-ethylene-terephthalate(PET).AppliedPhysicsA:MaterialsScience&Processing.1991;**53**(4):330-331

[8] Habibi Y. Key advances in the chemical modification of nanocelluloses. Chemical Society Reviews. 2014;**43**(5):1519-1542

[9] Xue C-H, Li Y-R, Zhang P, Ma J-Z, Jia S-T. Washable and wear-resistant superhydrophobic surfaces with self-cleaning property by chemical etching of fibers and hydrophobization. ACS Applied Materials & Interfaces. 2014;**6**(13):10153-10161

[10] Shi X, Xu L, Le TB, Zhou G, Zheng C, Tsuru K, et al. Partial oxidation of TiN coating by hydrothermal treatment and ozone treatment to improve its osteoconductivity. Materials Science and Engineering: C. 2016;**59**:542-548

[11] Puleo D, Kissling R, Sheu M-S. A technique to immobilize bioactive proteins, including bone morphogenetic protein-4 (BMP-4), on titanium alloy. Biomaterials. 2002;**23**(9):2079-2087

[12] Biederman H. Plasma Polymer Films. London: World Scientific; 2004

[13] Gogolewski S, Mainil-Varlet P, Dillon J. Sterility, mechanical properties, and molecular stability of polylactide internal-fixation devices treated with low-temperature plasmas. Journal of Biomedical Materials Research Part A. 1996;**32**(2):227-235

[14] Chen TF, Siow KS, Ng PY, Majlis BY. Enhancing the biocompatibility of the polyurethane methacrylate and off-stoichiometry thiol-ene polymers by argon and nitrogen plasma treatment. Materials Science and Engineering: C. 2017;**79**:613-621

[15] Wolf RA. Atmospheric Pressure Plasma for Surface Modification. Hoboken, New Jersey, United States: John Wiley & Sons; 2012

[16] Christophorou LG, Olthoff JK. Fundamental Electron Interactions with Plasma Processing Gase. Berlin, Germany: Springer Science & Business Media; 2012

[17] PL_INTL. Plasmas international. Perspectives on plasmas. 2004. Available at www.plasmas.org/what-are-plasmas.htm [Accessed on 2 August 2017]

[18] De Geyter N. Plasma Modification of Polymer Surfaces in the Subatmospheric Pressure Range; 2007-2008

[19] Yasuda HK. Fundamental aspects of ionized gas. In: Plasma Polymerization. Cambridge, Massachusetts, United States: Academic Press; 2012

[20] Alves CM, Yang Y, Carnes D, Ong J, Sylvia V, Dean D, et al. Modulating bone cells response onto starch-based biomaterials by surface plasma treatment and protein adsorption. Biomaterials. 2007;**28**(2):307-315

[21] Novotná Z, Rimpelová S, Juřík P, Veselý M, Kolská Z, Hubáček T, et al. The interplay of plasma treatment and gold coating and ultra-high molecular weight polyethylene: On the cytocompatibility. Materials Science and Engineering: C. 2017;**71**:125-131

[22] Lee JH, Park JW, Lee HB. Cell adhesion and growth on polymer surfaces with hydroxyl groups prepared by water vapour plasma treatment. Biomaterials. 1991;**12**(5):443-448

[23] Morent R, De Geyter N, Desmet T, Dubruel P, Leys C. Plasma surface modification of biodegradable polymers: A review. Plasma Processes and Polymers. 2011;**8**(3):171-190

[24] Desmet T, Morent R, Geyter ND, Leys C, Schacht E, Dubruel P. Nonthermal plasma technology as a versatile strategy for polymeric biomaterials surface modification: A review. Biomacromolecules. 2009;**10**(9):2351-2378

[25] De Geyter N, Morent R, Leys C, Gengembre L, Payen E. Treatment of polymer films with a dielectric barrier discharge in air, helium and argon at medium pressure. Surface and Coatings Technology. 2007;**201**(16):7066-7075

[26] Van Dyke LS, Brumlik CJ, Liang W, Lei J, Martin CR, Yu Z, et al. Modification of fluoropolymer surfaces with electronically conductive polymers. Synthetic Metals. 1994;**62**(1):75-81

[27] Os MT. Surface Modification by Plasma Polymerization: Film Deposition, Tailoring of Surface Properties and Biocompatibility. Universiteit Twente; 2000

[28] Goodman J. The formation of thin polymer films in the gas discharge. Journal of Polymer Science. 1960;**44**(144):551-552

[29] Stuart M. Dielectric properties of cross-linked polystyrene film formed in the glow discharge. Nature. 1963;**199**:59-60

[30] Bradley A, Hammes JP. Electrical properties of thin organic films. Journal of The Electrochemical Society. 1963;**110**(1):15-22

[31] Yasuda H. Plasma Polymerization. London, UK: Academic Press; 1985

[32] Bax DV, McKenzie DR, Weiss AS, Bilek MM. The linker-free covalent attachment of collagen to plasma immersion ion implantation treated polytetrafluoroethylene and subsequent cell-binding activity. Biomaterials. 2010;**31**(9):2526-2534

[33] Bax DV, Wang Y, Li Z, Maitz PK, McKenzie DR, Bilek MM, et al. Binding of the cell adhesive protein tropoelastin to PTFE through plasma immersion ion implantation treatment. Biomaterials. 2011;**32**(22):5100-5111

[34] Gan B, Bilek M, Kondyurin A, Mizuno K, McKenzie D. Etching and structural changes in nitrogen plasma immersion ion implanted polystyrene films. Nuclear Instruments and Methods in Physics Research Section B: Beam Interactions with Materials and Atoms. 2006;**247**(2):254-260

[35] Yasuda H, Hsu T. Plasma polymerization investigated by the comparison of hydrocarbons and perfluorocarbons. Surface Science. 1978;**76**(1):232-241

[36] Chiper A, Borcia G. Argon versus helium dielectric barrier discharge for surface modification of polypropylene and poly (methyl methacrylate) films. Plasma Chemistry and Plasma Processing. 2013;**33**(3):553-568

[37] Jacobs T, Morent R, De Geyter N, Dubruel P, Leys C. Plasma surface modification of biomedical polymers: Influence on cell-material interaction. Plasma Chemistry and Plasma Processing. 2012;**32**(5):1039-1073

[38] Liu Y, Su C, Ren X, Fan C, Zhou W, Wang F, et al. Experimental study on surface modification of PET films under bipolar nanosecond-pulse dielectric barrier discharge in atmospheric air. Applied Surface Science. 2014;**313**:53-59

[39] Inagaki N. Plasma Surface Modification and Plasma Polymerization. Boca Raton, Florida, United States: CRC Press; 1996

[40] Ponsonnet L, Reybier K, Jaffrezic N, Comte V, Lagneau C, Lissac M, et al. Relationship between surface properties (roughness, wettability) of titanium and titanium alloys and cell behaviour. Materials Science and Engineering: C. 2003;**23**(4):551-560

[41] Pandiyaraj KN, Kumar AA, RamKumar M, Deshmukh R, Bendavid A, P-G S, et al. Effect of cold atmospheric pressure plasma gas composition on the surface and cyto-compatible properties of low density polyethylene (LDPE) films. Current Applied Physics. 2016; **16**(7):784-792

[42] Gupta AK, Gupta M, Yarwood SJ, Curtis AS. Effect of cellular uptake of gelatin nanoparticles on adhesion, morphology and cytoskeleton organisation of human fibroblasts. Journal of Controlled Release. 2004;**95**(2):197-207

[43] Aziz G, Cools P, De Geyter N, Declercq H, Cornelissen R, Morent R. Dielectric barrier discharge plasma treatment of ultrahigh molecular weight polyethylene in different discharge atmospheres at medium pressure: A cell-biomaterial interface study. Biointerphases. 2015;**10**(2):029502

[44] Meyer-Plath A, Schröder K, Finke B, Ohl A. Current trends in biomaterial surface functionalization—Nitrogen-containing plasma assisted processes with enhanced selectivity. Vacuum. 2003;**71**(3):391-406

[45] Finke B, Luethen F, Schroeder K, Mueller PD, Bergemann C, Frant M, et al. The effect of positively charged plasma polymerization on initial osteoblastic focal adhesion on titanium surfaces. Biomaterials. 2007;**28**(30):4521-4534

[46] Robinson DE, Marson A, Short RD, Buttle DJ, Day AJ, Parry KL, et al. Surface gradient of functional heparin. Advanced Materials. 2008;**20**(6):1166-1169

[47] Xu J, Gleason KK. Conformal, amine-functionalized thin films by initiated chemical vapor deposition (iCVD) for hydrolytically stable microfluidic devices. Chemistry of Materials. 2010;**22**(5):1732-1738

[48] Detomaso L, Gristina R, Senesi GS, d'Agostino R, Favia P. Stable plasma-deposited acrylic acid surfaces for cell culture applications. Biomaterials. 2005;**26**(18):3831-3841

[49] Zhang Z, Chen Q, Knoll W, Foerch R, Holcomb R, Roitman D. Plasma polymer film structure and DNA probe immobilization. Macromolecules. 2003;**36**(20):7689-7694

[50] Zhang Z, Knoll W, Förch R. Amino-functionalized plasma polymer films for DNA immobilization and hybridization. Surface and Coating Technology. 2005;**200**(1):993-995

[51] Muguruma H, Hiratsuka A, Karube I. Thin-film glucose biosensor based on plasma-polymerized film: Simple design for mass production. Analytical Chemistry. 2000;**72**(11): 2671-2675

[52] Gupta B, Plummer C, Bisson I, Frey P, Hilborn J. Plasma-induced graft polymerization of acrylic acid onto poly (ethylene terephthalate) films: Characterization and human smooth muscle cell growth on grafted films. Biomaterials. 2002;**23**(3):863-871

[53] Yuan S, Szakalas-Gratzl G, Ziats NP, Jacobsen DW, Kottke-Marchant K, Marchant RE. Immobilization of high-affinity heparin oligosaccharides to radiofrequency plasma-modified polyethylene. Journal of Biomedical Materials Research Part A. 1993;**27**(6):811-819

[54] Siow KS, Britcher L, Kumar S, Griesser HJ. Plasma methods for the generation of chemically reactive surfaces for biomolecule immobilization and cell colonization–a review. Plasma Processes and Polymers. 2006;**3**(6-7):392-418

[55] Deng X, Nikiforov AY, Coenye T, Cools P, Aziz G, Morent R, et al. Antimicrobial nano-silver non-woven polyethylene terephthalate fabric via an atmospheric pressure plasma deposition process. Scientific Reports. 2015;**5**:10138

[56] Kumar V, Jolivalt C, Pulpytel J, Jafari R, Arefi-Khonsari F. Development of silver nanoparticle loaded antibacterial polymer mesh using plasma polymerization process. Journal of Biomedical Materials Research Part A. 2013;**101**(4):1121-1132

[57] Degoutin S, Jimenez M, Casetta M, Bellayer S, Chai F, Blanchemain N, et al. Anticoagulant and antimicrobial finishing of non-woven polypropylene textiles. Biomedical Materials. 2012;**7**(3):035001

[58] Detomaso L, Gristina R, d'Agostino R, Senesi GS, Favia P. Plasma deposited acrylic acid coatings: Surface characterization and attachment of 3T3 murine fibroblast cell lines. Surface and Coatings Technology. 2005;**200**(1):1022-1025

[59] Majani R, Zelzer M, Gadegaard N, Rose FR, Alexander MR. Preparation of Caco-2 cell sheets using plasma polymerised acrylic acid as a weak boundary layer. Biomaterials. 2010;**31**(26):6764-6771

[60] Senesi GS, D'Aloia E, Gristina R, Favia P, d'Agostino R. Surface characterization of plasma deposited nano-structured fluorocarbon coatings for promoting in vitro cell growth. Surface Science. 2007;**601**(4):1019-1025

[61] Gristina R, D'Aloia E, Senesi GS, Milella A, Nardulli M, Sardella E, et al. Increasing cell adhesion on plasma deposited fluorocarbon coatings by changing the surface topography. Journal of Biomedical Materials Research Part B: Applied Biomaterials. 2009;**88**(1): 139-149

[62] Dhayal M, Cho S-I. Leukemia cells interaction with plasma-polymerized acrylic acid coatings. Vacuum. 2006;**80**(6):636-642

[63] Intranuovo F, Sardella E, Gristina R, Nardulli M, White L, Howard D, et al. PE-CVD processes improve cell affinity of polymer scaffolds for tissue engineering. Surface and Coatings Technology. 2011;**205**:S548-SS51

[64] Finke B, Hempel F, Testrich H, Artemenko A, Rebl H, Kylián O, et al. Plasma processes for cell-adhesive titanium surfaces based on nitrogen-containing coatings. Surface and Coating Technology. 2011;**205**:S520-S5S4

[65] Daw R, Candan S, Beck AJ, Devlin AJ, Brook IM, MacNeil S, et al. Plasma copolymer surfaces of acrylic acid/1, 7 octadiene: Surface characterisation and the attachment of ROS 17/2.8 osteoblast-like cells. Biomaterials. 1998;**19**(19):1717-1725

[66] Buttiglione M, Vitiello F, Sardella E, Petrone L, Nardulli M, Favia P, et al. Behaviour of SH-SY5Y neuroblastoma cell line grown in different media and on different chemically modified substrates. Biomaterials. 2007;**28**(19):2932-2945

[67] Zheng Y, Xiong C, Wang Z, Li X, Zhang L. A combination of CO 2 laser and plasma surface modification of poly (etheretherketone) to enhance osteoblast response. Applied Surface Science. 2015;**344**:79-88

[68] Pezzatini S, Morbidelli L, Gristina R, Favia P, Ziche M. A nanoscale fluorocarbon coating on PET surfaces improves the adhesion and growth of cultured coronary endothelial cells. Nanotechnology. 2008;**19**(27):275101

[69] Myung SW, Ko YM, Kim BH. Protein adsorption and cell adhesion on three-dimensional polycaprolactone scaffolds with respect to plasma modification by etching and deposition techniques. Japanese Journal of Applied Physics. 2014;**53**(11S):11RB01

[70] Hamerli P, Weigel T, Groth T, Paul D. Surface properties of and cell adhesion onto allylamine-plasma-coated polyethylenterephtalat membranes. Biomaterials. 2003;**24**(22): 3989-3999

[71] Mitchell S, Davidson M, Emmison N, Bradley R. Isopropyl alcohol plasma modification of polystyrene surfaces to influence cell attachment behaviour. Surface Science. 2004; **561**(1):110-120

[72] Muir BW, Nelson A, Fairbrother A, Fong C, Hartley PG, James M, et al. A comparative X-ray and neutron reflectometry study of plasma polymer films containing reactive amines. Plasma Processes and Polymers. 2007;**4**(4):433-444

[73] Zhang Z, Chen Q, Knoll W, Förch R. Effect of aqueous solution on functional plasma polymerized films. Surface and Coating Technology. 2003;**174**:588-590

[74] Jeon H, Wyatt J, Harper-Nixon D, Weinkauf D. Characterization of thin polymer-like films formed by plasma polymerization of methylmethacrylate: A neutron reflectivity study. Journal of Polymer Science Part B: Polymer Physics. 2004;**42**(13):2522-2530

[75] Tamirisa PA, Hess DW. Water and moisture uptake by plasma polymerized thermoresponsive hydrogel films. Macromolecules. 2006;**39**(20):7092-7097

[76] Vasilev K, Britcher L, Casanal A, Griesser HJ. Solvent-induced porosity in ultrathin amine plasma polymer coatings. The Journal of Physical Chemistry B. 2008;**112**(35):10915-10921

[77] Förch R, Zhang Z, Knoll W. Soft plasma treated surfaces: Tailoring of structure and properties for biomaterial applications. Plasma Processes and Polymers. 2005;**2**(5):351-372

[78] Tarasova A, Hamilton-Brown P, Gengenbach T, Griesser HJ, Meagher L. Colloid probe AFM and XPS study of time-dependent aging of amine plasma polymer coatings in aqueous media. Plasma Processes and Polymers. 2008;**5**(2):175-185

[79] Aziz G, Thukkaram M, De Geyter N, Morent R. Plasma parameters effects on the properties, aging and stability behaviors of allylamine plasma coated ultra-high molecular weight polyethylene (UHMWPE) films. Applied Surface Science. 2017;**409**:381-395

[80] Gengenbach TR, Vasic ZR, Chatelier RC, Griesser HJ. A multi-technique study of the spontaneous oxidation of N-hexane plasma polymers. Journal of Polymer Science Part A: Polymer Chemistry. 1994;**32**(8):1399-1414

[81] Gengenbach TR, Griesser HJ. Compositional changes in plasma-deposited fluorocarbon films during ageing. Surface and Interface Analysis. 1998;**26**(7):498-511

[82] Gengenbach TR, Griesser HJ. Aging of 1, 3-diaminopropane plasma-deposited polymer films: Mechanisms and reaction pathways. Journal of Polymer Science Part A: Polymer Chemistry. 1999;**37**(13):2191-2206

[83] Gengenbach TR, Griesser HJ. Post-deposition ageing reactions differ markedly between plasma polymers deposited from siloxane and silazane monomers. Polymer. 1999;**40**(18): 5079-5094

[84] Goreham RV, Short RD, Vasilev K. Method for the generation of surface-bound nanoparticle density gradients. The Journal of Physical Chemistry C. 2011;**115**(8):3429-3433

[85] Menzies DJ, Cowie B, Fong C, Forsythe JS, Gengenbach TR, McLean KM, et al. One-step method for generating PEG-like plasma polymer gradients: Chemical characterization and analysis of protein interactions. Langmuir. 2010;**26**(17):13987-13994

[86] Goreham RV, Mierczynska A, Pierce M, Short RD, Taheri S, Bachhuka A, et al. A substrate independent approach for generation of surface gradients. Thin Solid Films. 2013; **528**:106-110

[87] Harding F, Goreham R, Short R, Vasilev K, Voelcker NH. Surface bound amine functional group density influences embryonic stem cell maintenance. Advanced Healthcare Materials. 2013;**2**(4):585-590

[88] Zelzer M, Alexander MR, Russell NA. Hippocampal cell response to substrates with surface chemistry gradients. Acta Biomaterialia. 2011;**7**(12):4120-4130

[89] Zelzer M, Scurr D, Abdullah B, Urquhart AJ, Gadegaard N, Bradley JW, et al. Influence of the plasma sheath on plasma polymer deposition in advance of a mask and down pores. The Journal of Physical Chemistry B. 2009;**113**(25):8487-8494

[90] Mangindaan D, Kuo W-H, Wang M-J. Two-dimensional amine-functionality gradient by plasma polymerization. Biochemical Engineering Journal. 2013;**78**:198-204

[91] Parry KL, Shard A, Short R, White R, Whittle J, Wright A. ARXPS characterisation of plasma polymerised surface chemical gradients. Surface and Interface Analysis. 2006;**38**(11): 1497-1504

[92] Pitt WG. Fabrication of a continuous wettability gradient by radio frequency plasma discharge. Journal of Colloid and Interface Science. 1989;**133**(1):223-227

[93] Whittle JD, Barton D, Alexander MR, Short RD. A method for the deposition of controllable chemical gradients. Chemical Communications. 2003;**14**:1766-1767

[94] Vasilev K, Mierczynska A, Hook AL, Chan J, Voelcker NH, Short RD. Creating gradients of two proteins by differential passive adsorption onto a PEG-density gradient. Biomaterials. 2010;**31**(3):392-397

[95] Ghobeira R, Philips C, Declercq H, Cools P, De Geyter N, Cornelissen R, et al. Effects of different sterilization methods on the physico-chemical and bioresponsive properties of plasma-treated polycaprolactone films. Biomedical Materials. 2017;**12**(1):015017

[96] Walker RA, Cunliffe VT, Whittle JD, Steele DA, Short RD. Submillimeter-scale surface gradients of immobilized protein ligands. Langmuir. 2009;**25**(8):4243-4246

[97] Wells N, Baxter MA, Turnbull JE, Murray P, Edgar D, Parry KL, et al. The geometric control of E14 and R1 mouse embryonic stem cell pluripotency by plasma polymer surface chemical gradients. Biomaterials. 2009;**30**(6):1066-1070

[98] Harding FJ, Clements LR, Short RD, Thissen H, Voelcker NH. Assessing embryonic stem cell response to surface chemistry using plasma polymer gradients. Acta Biomaterialia. 2012;**8**(5):1739-1748

[99] Wang P-Y, Clements LR, Thissen H, Tsai W-B, Voelcker NH. Screening rat mesenchymal stem cell attachment and differentiation on surface chemistries using plasma polymer gradients. Acta Biomaterialia. 2015;**11**:58-67

[100] Mangindaan D, Kuo WH, Kurniawan H, Wang MJ. Creation of biofunctionalized plasma polymerized allylamine gradients. Journal of Polymer Science Part B: Polymer Physics. 2013;**51**(18):1361-1367

[101] Barry JJ, Howard D, Shakesheff KM, Howdle SM, Alexander MR. Using a core–sheath distribution of surface chemistry through 3D tissue engineering scaffolds to control cell ingress. Advanced Materials. 2006;**18**(11):1406-1410

[102] Zelzer M, Majani R, Bradley JW, Rose FR, Davies MC, Alexander MR. Investigation of cell–surface interactions using chemical gradients formed from plasma polymers. Biomaterials. 2008;**29**(2):172-184

[103] Robinson DE, Buttle DJ, Whittle JD, Parry KL, Short RD, Steele DA. The substrate and composition dependence of plasma polymer stability. Plasma Processes and Polymers. 2010;**7**(2):102-106

[104] Delalat B, Goreham RV, Vasilev K, Harding FJ, Voelcker NH. Subtle changes in surface chemistry affect embryoid body cell differentiation: Lessons learnt from surface bound amine density gradients. Tissue Engineering Parts A. 2014;**20**(11-12):1715-1725

[105] Liu X, Shi S, Feng Q, Bachhuka A, He W, Huang Q, et al. Surface chemical gradient affects the differentiation of human adipose-derived stem cells via ERK1/2 signaling pathway. ACS Applied Materials & Interfaces. 2015;**7**(33):18473-18482

Chapter 5

Sorption Properties of Clay and Pectin-Containing Hydrogels

Mayya V. Uspenskaya, Vera E. Sitnikova,
Michael A. Dovbeta, Roman O. Olekhnovich and
Igor Yu. Denisyuk

Additional information is available at the end of the chapter

http://dx.doi.org/10.5772/intechopen.71190

Abstract

As is known, polymeric polyelectrolyte hydrogels are superabsorbents that are capable of absorbing moisture in amounts many times greater than their own mass. Numerous studies have shown that besides water absorption and retention, they can also be effectively used as sorbents to purify water from heavy metals. In many works, attempts are made to improve the sorption properties of polyelectrolyte hydrogels by creating polymer composites based on them. Organic/inorganic composite materials frequently exhibited desired hybrid performance superior to their individual components and cost-efficient characteristics. The composites derived from natural polysaccharides and inorganic clay minerals are of special interest by virtue of their unique commercial and environmental advantages, which means that the design and development of environmentally friendly superabsorbents, introducing natural ingredients, have long been necessary. In this paper, we consider polymer hydrogels based on a copolymer of acrylic acid and acrylamide filled with pectin and bentonite. The aim of this study is to investigate the influence of chemical conditions on hydrogels and their composites, kinetic, and absorption behavior toward metal ions in the presence of the chelating agent. In this chapter, an investigation of the kinetic patterns of swelling, deswelling, and sorption of the hydrogels and their composites will be presented.

Keywords: hydrogel, semi-interpenetrating networks, bentonite, pectin

1. Introduction

The term "hydrogel" is ambiguous, and it must first be clarified. We will call gel, a polymer network swollen in a solvent—a set of a large number of polymer chains chemically (or physically) sewn together. More strictly, the polymer gel is a system consisting of at least two components, one of which is a mesh polymer and the other is a liquid present in a palpable amount.

Polymer gels can almost completely (by 99% or more) consist of a liquid and be very soft materials. Despite this, they have the inherent ability of solid bodies to maintain their shape. This is due to the fact that the polymer mesh that forms part of the gel plays the role of a framework that provides rigidity (elasticity) of the entire system, i.e., it does not allow it to flow under the action of a deforming force (if the force is not too great and does not last too long).

1.1. Composite polymer hydrogels

The combination of the properties of hydrogels inherent in solids predetermines a wide range of applications from technical spheres (sorbents, gas separating, and ion exchange membranes [1–3]) to the food industry and medicine (food structure, drug carriers, artificial substitutes for biological tissues, materials for soft and intraocular lenses, etc. [4–6]). The emergence of new fields of application of polymer hydrogels puts forward new requirements to their properties.

These goals can be achieved by obtaining fundamentally new materials—composite hydrogels containing at least two components, each of which performs certain functions. It is obvious that the characteristics of the composite hydrogel are due not only to the physicochemical properties of the individual components, but also to the structure of the material. Given the limited thermodynamic compatibility of the polymers, a variety of hydrogel structures are possible, from complete stratification of the polymer phases to the formation of matrix-nanoscale structures or the formation of structures in which both polymer phases are continuous.

In most cases, composite hydrogels are biphasic systems. The interphase boundary in such materials is not always clearly expressed. It can be a transition layer in which a gradual change in properties occurs (transition from the properties of phase 1 to the properties of phase 2). At least one of phases must be a polymer hydrogel. The hydrogel can be either a synthetic or a natural polymer. The second phase can be a synthetic hydrogel of synthetic or natural origin, a hydrophobic polymer and an inorganic substance. In accordance with the foregoing, it is possible to propose a classification that divides composite hydrogels into three groups [7]:

1. Hydrogels consisting of two hydrophilic polymers, each of which is capable of forming an individual polymer hydrogel.
2. Hydrogel, including hydrophilic and hydrophobic polymers.
3. A polymeric hydrogel containing an inorganic phase.

The nature of the interactions between the components may be due to covalent bonds in the block and graft copolymers, the formation of interpolymer complexes due to the formation of hydrogen bonds, donor-acceptor, ionic and hydrophobic interactions of functional groups, and the engagement of macromolecular chains in interpenetrating and semi-interpenetrating polymer networks.

The compositional polymeric materials, consisting of hydrophilic polymers, have most application in last time. The presence of charged polar groups in hydrophilic polymers lids to communication with solution particles at help of the formation of intermolecular hydrogen and ionic bonds between molecules. Under conditions of corresponding implementation of functional groups, covalent bonds can exist in composite hydrogels. The hydrogels form by the type of interpenetrating and semi-interpenetrating polymer networks usually. In this case, the intermolecular bonds are caused by physical interactions between polar groups.

Inorganic components in composition hydrogels are introduced either to modify the properties of conventional polymer hydrogels (changing mechanical properties, increasing the sensitivity of hydrogels to thermal effects, changing pH, etc.) or to impart new properties not typical for hydrogels (magnetic characteristics and antibacterial properties). Oxides, various clays, carbon materials, water-insoluble inorganic salts, and metals are most often used in organo-inorganic composite hydrogels.

Methods for the preparation of organo-inorganic hydrogels can be divided into two groups:

1. Mixing of inorganic additives in the form of nano- or microparticles with a solution of a water-soluble polymer or monomers, followed by their polymerization. In such cases, the filler particles often serve as physical crosslinking centers.
2. Formation of the inorganic phase as a result of sol-gel process, which involves the introduction of monomers or polymer, precursors of the inorganic component into the solution, and its subsequent conversion in the course of various chemical reactions into solid water-insoluble particles. The final step is the polymerization of the monomers or, if required, crosslinking of the polymer.

1.2. Swelling and collapse of polyelectrolyte gels

Polymer gel, placed in a solvent, changes its volume, i.e., swells or contracts, appropriately absorbing or releasing the solvent, until it reaches an equilibrium swelling. The equilibrium degree of swelling of the gel, determined by the amount of solvent in it, depends both on the properties of the gel and on the properties of the solvent. The nature of this relationship was established for the first time in theoretical studies carried out by Flory and Rener [8, 9] and Kachalsky [10, 11]. According to Flori-Rener's postulate, the equilibrium of free swelling of the polymer network is determined by the balance between the mesh expanding the osmotic pressure and the elastic stress that arises in it. In 1977, Peppas and Merrill modified the Flory-Rehner theory in application to the production of hydrogels from polymer solutions. Due to elastic forces, the presence of water affects the change in the chemical potential within the system [12].

Of all polymer gels, the most interesting are gels based on chains containing charged units, polyelectrolyte gels. Since the macroscopic sample of the gel must be electrically neutral, the charge of the polymer chains of the gel must be compensated by the opposite charge of the low molecular weight counterions floating in the solvent in which the gel swells. When the gel swells in a large volume of water, the counterions should be advantageous to leave the gel and go to the external solution, which would lead to a significant gain in their translational entropy. However, this does not occur, since this leads to a violation of the electroneutrality condition of the gel sample. Counterions are forced to stay inside the gel and create there a bursting osmotic pressure. This osmotic pressure is responsible for the two most important effects associated with polyelectrolyte gels swelling in water.

First, a simple theory shows [13, 14] that the effect of the expanding osmotic pressure is very strong, it leads to a significant swelling of the gel in the water. Therefore, polyelectrolyte gels are used as superabsorbents of water.

Second, the superstrong swelling of polyelectrolyte gels in water leads to the fact that their concentration decreases extremely sharply with a deterioration in the quality of the solvent. The volume of the gel may decrease by a factor of thousands. This phenomenon, called the collapse of gels, was first predicted theoretically in [15] and was experimentally found in [16]. It is associated with the transition of the tangle-globule in the chains constituting the polymer gel. As a result, the gel sample collapses as a whole. In this case, the higher the degree of gel charge, the more sharply the collapse occurs [13]. The theory of collapse of gels developed in [13, 17, 18] shows that this is due to the fact that the collapsed phase is stabilized by the forces of attraction of uncharged links and the volume of the gel in this case depends little on the degree of charge, whereas the volume of the swollen gel is substantially increases with an increase in the degree of charge due to the expanding osmotic pressure of the counterions.

As discussed earlier in the works [19, 20], the equilibrium swelling of ionic hydrogels depends on the network structure, degree of crosslinking, hydrophilicity, and ionization of the functional groups. The major factor contributing to the swelling of ionic networks is the ionization of the network.

2. Experimental

In this chapter, the sorption properties of two different types of hydrogel composites will be compared. The first type is a clay-containing hydrogel composite—a copolymer of acrylic acid and acrylamide, filled with bentonite. The second type is a hydrogel filled with polysaccharide—pectin. Pectin is a naturally occurring biopolymer that is finding increasing applications in the pharmaceutical and biotechnology industry. The combination of the hydrophilic acrylic polymer properties with the biodegradable character of pectin-based blends, can lead to interesting hydrogels with potential applications as biomaterials exhibiting different properties depending on the composition and on the type of interactions within the network, attending to chemical crosslinking and hydrogen bonding interactions. In addition, a hydrogel composite based on a copolymer of acrylic acid and acrylamide, filled with such a polysaccharide as pectin, is a hydrogel with a semi-interpenetrating network.

2.1. Material

Pectin (chemical grade, M_W 50,000) was purchased from Merck Chemical Co. (Germany). N,N′-methylene-bis-acrylamide (MBA), sodium hydroxide, ammonium persulfate, and bentonite were supplied by Sigma-Aldrich and were used without any further purification. Acrylic acid was distilled under reduced pressure before use. Acrylic acid (AAc) and acrylamide (AAm) were from Merck and were used without any further purification. All agents were of analytical grade quality.

2.2. Preparation of superabsorbent composites

Crosslinked acrylamide-sodium acrylate hydrogel and its composites were synthesized by free radical solution polymerization of AAm and AAc monomers in aqueous solution. A series of hydrogels were prepared by the following procedure.

Polymerization is carried out with constant stirring with a magnetic stirrer at a speed of 500 rev/min. According to the procedure, 10 ml of AAc monomer were dissolved in 3.5 ml distilled water. Acrylic acid was neutralized with 14 N aqueous solution of potassium hydroxide. The degree of neutralization is 0.8. Then 3 g of acrylamide monomer was added. To increase the crosslink density was added 0.001 g N,N-methylenebisacrylamide crosslinking agent. To initiate radical polymerization process a redox system consisting of 4 ml of a 2% ammonium persulfate aqueous solution and 4 ml of a 0.5% solution of TEMED are used. The polymerization was carried out at 35°C (**Figure 1**).

For the preparation of poly(AAm-co-AAc)/bentonite composite hydrogels, 1, 2, 3, or 4 mas.% bentonite was added into solution of sodium acrylate (AAcNa), AAm, and MBA, and was stirred for 10 min. Then, an oxidation–reduction system was added.

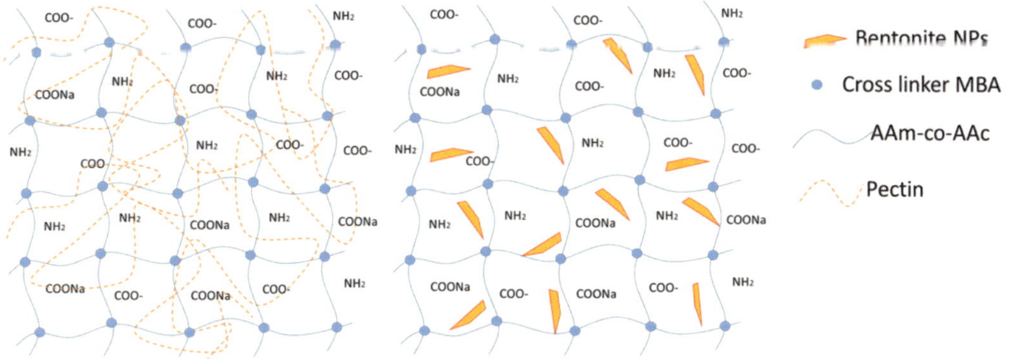

Figure 1. Schematic representation of a hydrogel with semi-interpenetrating networks—poly(AAm-co-AAc)/pectin and a dispersion-filled hydrogel composite—poly(AAm-co-AAc)/bentonite.

Poly(AAm-co-AAc)/pectin composite semi-IPNs were prepared using the same preparation method. However, the pectin powder was dissolved in a solution of sodium acrylate, stirred with a magnetic stirrer for 10 min, then acrylamide, a crosslinking agent and a redox system were added. To prepare highly swollen poly(AAm-co-AAc)/pectin (containing different contents of pectin) semi-IPNs, same method was used as mentioned above with the addition of 1, 2, 3, or 4 mas.% of pectin to solution of sodium acrylate.

3. Results and discussions

3.1. Characterization of hydrogels

IR spectra confirm the formation of a copolymer of acrylamide and acrylic acid, as seen from the bands that appeared in the range of 3100–3500 cm^{-1} (O—H and N—H stretching) (**Figure 2**). Absorption bands located in the region 3350–3330 and 3200–3185 cm^{-1} corresponds to the asymmetrical and symmetrical stretching $—NH_2$ of acrylamide [21].

On the other hand, the broad absorption band of 3400–2950 cm^{-1} may be attributed to the —OH of the carboxyl group. Absorption bands in the regions of 2950–2940 and 2915–2900 cm^{-1} correspond to asymmetric and symmetric stretching of the $—CH_2$ groups. In addition, the absorption band at 2790–2770 cm^{-1} is characteristic stretching —CH group of the polymer chain. The stretching of the —C= group of acrylamide and acrylic acid in the frequency of 1653–1645 cm^{-1} appear in all spectra of the hydrogel composites. The typical band, which

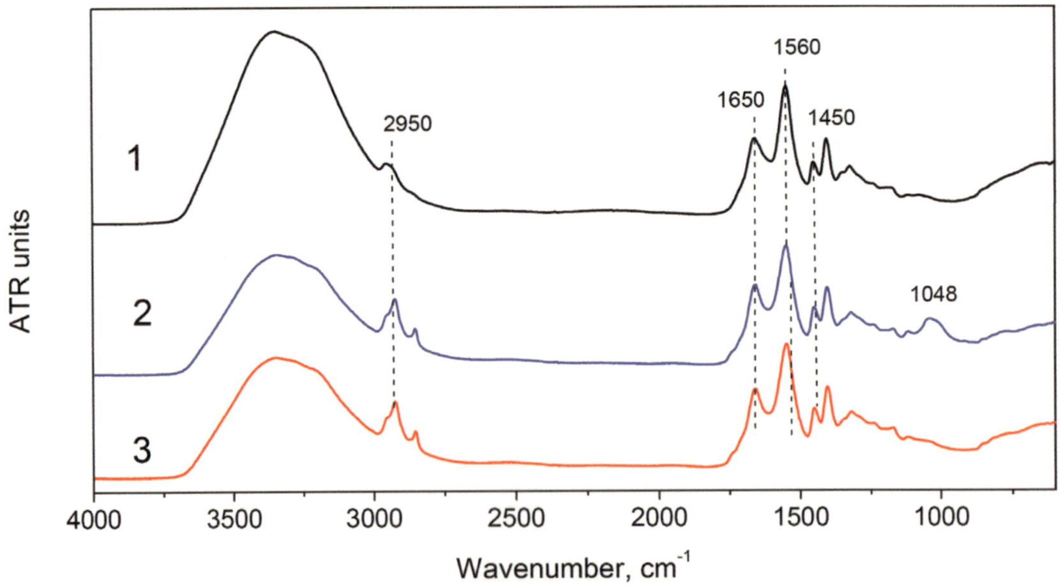

Figure 2. ATR FTIR spectra of poly(AAm-co-AAc)/pectin (1), poly(AAm-co-AAc)/bentonite composites (2), and poly(AAm-co-AAc) hydrogel (3).

appeared in the range of 1620–1600 cm^{-1}, results to deformation vibrations group —NH_2 acrylamide. The absorption band in the 1455–1445 cm^{-1} belongs to the stretching vibrations group —CH_2. The absorption band, which is located in the region 1450–1410 cm^{-1}, is a characteristic for the C—H stretching. The absorption bands at 1560 and 1406–1410 cm^{-1} results to the symmetric and asymmetric stretching of —COO-acrylate (acrylic acid neutralized with NaOH). The presence of absorbed water can be seen through the band at 3300 cm^{-1} in all spectra of the samples.

The absorption band 987 cm^{-1} of Si—O—Si of bentonite in the spectra composites is shifted from the frequency of 1048 cm^{-1}. This indicates that the composite components interact with each other by complexation to form homogeneous gels. From the FTIR it is clear that there is no significant shift in major peaks, which indicates that there is no chemical interaction between the polymer and the pectin used.

As can be seen from the SEM image (**Figure 3**), pectin is evenly distributed throughout the volume of the hydrogel. The structure of composite existing pectinate scaffold resulted in a double network architecture, where filamentous polyAA-co-AAm networks penetrated through pores of the pectin network.

3.2. Study of water absorption properties

3.2.1. Measurement of swelling

Swelling kinetic experiments were carried out by immersing a known amount of the dried hydrogels with 100 mL of distilled water in a constant temperature at 25°C. Gravimetrical measurement method was used to measure the swelling rate of the hydrogels. The water absorption amount Q (g/g) was calculated as follows:

$$Q = \frac{m - m_0(1-\gamma)}{m_0(1-\gamma)}, \quad (1)$$

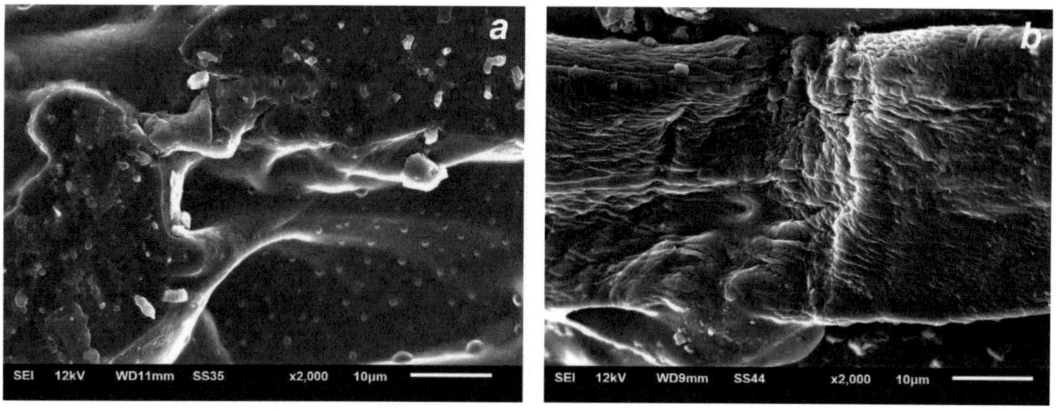

Figure 3. SEM image at 20.0 kV, showing surface structures of poly(AAm-co-AAc)/bentonite 3% (a) and poly(AAm-co-AAc)/pectin 3% (b).

where m_0 and m (g) are the weights of the dry and swollen sample, respectively, and γ is the water content of the hydrogel. Q was calculated as grams of water per gram of dry sample.

The hydrophilic properties of poly(AAm-co-AAc) and its composites hydrogels filled with bentonite and pectin were investigated by measuring their water uptake. Water uptake values were obtained by the mass ratio of the swollen hydrogel to dried hydrogel. **Figure 4** shows the dependences of water uptake as a function of the immersion time of the hydrogels swelled in distilled water.

Polymer hydrogel swells when it is brought into contact with a distilled water and salt aqueous solution. Swelling of the polymer hydrogel continues until the forces due to swelling of the polymer balance the osmotic pressure, driving the solvent into the swollen polymer. The swelling process of hydrogels is a complicated phenomenon and involves three successive steps: (i) the diffusion of water molecules into the polymer network, (ii) the relaxation of hydrated polymer chains, and (iii) the expansion of the polymer network into the surrounding aqueous solution [22].

Figure 4 presents the dynamic swelling data for poly(AAm-co-AAc)/pectin and poly(AAm-co-AAc)/bentonite in distilled water.

For each samples, the swelling ratio increases with time until a certain point, when it becomes constant, i.e., the equilibrium state is reached. At the beginning, the swelling is very fast. Besides, all hydrogels exhibit a salt-responsive swelling. Values of Q_{eq} decrease with increasing salt concentration. Moreover, the results clearly reveal that all hydrogels undergo drastic mass and/or volume change in the definite concentration range.

Poly(AAm-co-AAc)/pectin and poly(AAm-co-AAc)/bentonite hydrogel composites differ in their molecular structure (hydrogel containing dispersed filler particles and a mixture of linear and crosslinked polymers—semi-interpenetrating networks), however, the degree of crosslinking, the content of ionizable groups, and the supramolecular structure (degree of crystallinity)

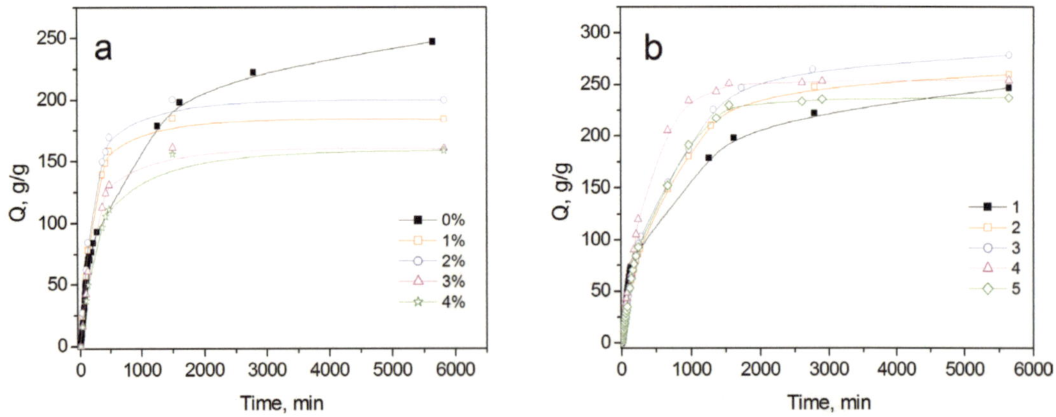

Figure 4. Swelling curves of poly(AAm-co-AAc)/pectin semi-IPN hydrogels (a) with 0% (1), 1% (2), 2% (3), 3% (4), 4% (5) pectin and poly(AAm-co-AAc)/bentonite (b) with 0% (1), 1% (2), 2% (3), 3% (4), 4% (5), 5% (6) bentonite in water.

are the same. Bearing in mind these differences in molecular and supermolecular structure of dispersed composite and semi-IPN and all factors affecting the swelling of ionic hydrogels, we could look for some differences in poly(AAm-co-AAc)/pectin and poly(AAm-co-AAc)/bentonite swelling behavior.

It was shown that there was a decrease in the equilibrium swelling (Q_{eq}) of the semi-IPN systems when pectin was added to the hydrogel systems. Incorporation of pectin into the copolymer network leads to lower degrees of swelling. The reason of this is the polymeric structure of pectin. Here, it could be said that chains of pectin were placed in the crosslinked polymeric systems, instead of crosslinked AAm and AAc molecules. So, it was seen that there was a decrease in the value of the Q_{eq}, because of the decrease in the hydrophilic character at crosslinked polymeric systems. But there was generally an increase in the Q_{eq} of the hydrogel systems when bentonite was added to the hydrogel systems. It was seen that there was an increase in the value of the Q_{eq}, because of the increase in the hydrophilic character at crosslinked polymeric systems. However, an increase in the percentage of filling bentonite in the hydrogel composite leads to a deterioration in the swelling properties, since at high filler concentrations it acts as an additional crosslinking agent.

To characterize the effect of the swelling medium on the kinetics of water uptake of poly(AAm-co-AAc)/bentonite and poly(AAm-co-AAc)/pectin hydrogel composites, kinetic modeling was conducted on the basis of the Fickian diffusion law.

3.2.2. *Diffusion water*

Many mathematical models have been proposed to describe the kinetics of hydrogel swelling. Most dynamic hydrogel swelling models are based in some way on Fick's laws of diffusion. To determine the type of diffusion of water into hydrogels the simple and commonly used method, based on the power-law expression (Eq. (3)) was applied [23]:

$$\frac{Q_t}{Q_{eq}} = kt^n \tag{2}$$

where Q_t and Q_{eq} represent the amount of solvent diffused into the gel at time t and at infinite time (equilibrium state), respectively, k is a constant related to the structure of the network, and the exponent n is a number that determines the type of diffusion. This equation can be applied only to the initial stages of swelling, i.e., to a 60% increase in the mass of hydrogel ($Q_t/Q_{eq} \leq 0.6$; $\log(Q_t/Q_{eq}) \leq -0.22$).

The phenomenon of water sorption by hydrogels depends on the diffusion of water molecules into the gel matrix and subsequent relaxation of macromolecular chains of the hydrogel. The mechanism of water transport in a swelling hydrogel is greatly affected by many factors, such as equilibrium water content, chemical composition of the hydrogel, swelling rate, etc. [24]. In general, three models are used to describe transport phenomena into polymer networks [25].

According to the relative rates of diffusion (R_{diff}) and polymer relaxation (R_{relax}), three classes of diffusion can be distinguished. For a planar geometry, the value of (i) n = 0.5 indicates a Fickian diffusion mechanism (Case I) in which the rate of diffusion is much smaller than the rate of relaxation ($R_{diff} \ll R_{relax}$, system controlled by diffusion), (ii) n = 1.0 indicates Case II, where the diffusion process is much faster than the relaxation process ($R_{diff} \gg R_{relax}$, system controlled by relaxation), (iii) $0.5 < n < 1.0$ indicates non-Fickian (anomalous) diffusion mechanism, which describes those cases where the diffusion and relaxation rates are comparable ($R_{diff} \approx R_{relax}$). Occasionally, values of $n > 1$ have been observed, which are regarded as Super Case II kinetics. When the water penetration rate is much below the polymer chain relaxation rate, it is possible to record the n values below 0.5. This situation, which is classified also as Fickian diffusion, is called as "Less Fickian" behavior.

Calculation of the exponent n and constant k was achieved by plotting the data in log-log plots, according to Eq. (3) and estimating the obtained curves by linear functions

$$log\frac{Q_t}{Q_{eq}} = logk + n\,logt \tag{3}$$

The study of diffusion phenomena of water in hydrogels is of value in that it clarifies polymer behavior. **Table 1** shows that the number determining the type of diffusion (n) is over 0.50. Hence, the diffusion of water into the hydrogel systems is generally found to have a non-Fickian character. When the diffusion type is anomalous behavior, the relaxation and diffusion time are of the same order of magnitude. Thus, both diffusion and polymer relaxation control the overall rate of water uptake. However, poly(AAm-co-AAc) has the values of $n > 1$, which means a Super Case II kinetics, where the diffusion process is much faster than the relaxation process ($R_{diff} \gg R_{relax}$, system controlled by relaxation). Thus, the addition of bentonite and pectin to the hydrogel increases the rate of diffusion of water into the hydrogel composite, due to the increased hydrophilic properties of this composite. Moreover, pectin increases the rate of diffusion of water into the hydrogel to a greater extent.

3.2.3. *Swelling kinetics*

To describe the swelling kinetics of different hydrogel, three empirical models, namely, Peleg's, first-order, and Schott's second-order absorption kinetic model, are used.

Hydrogel	**k (min^{-1})**	**n**	**Mechanism**
Poly(AAm-co-AAc)	0.4449	1.0507	Super Case II transport
Poly(AAm-co-AAc)/bentonite 1%	0.0038	0.8480	Non-Fickian (anomalous) diffusion
Poly(AAm-co-AAc)/bentonite 2%	0.0029	0.8757	Non-Fickian diffusion
Poly(AAm-co-AAc)/bentonite 3%	0.0031	0.9217	Non-Fickian diffusion
Poly(AAm-co-AAc)/bentonite 4%	0.0025	0.9325	Non-Fickian diffusion
Poly(AAm-co-AAc)/pectin 1%	0.0078	0.8277	Non-Fickian diffusion
Poly(AAm-co-AAc)/pectin 2%	0.0088	0.8008	Non-Fickian diffusion
Poly(AAm-co-AAc)/pectin 3%	0.0044	0.9211	Non-Fickian diffusion
Poly(AAm-co-AAc)/pectin 4%	0.0057	0.8321	Non-Fickian diffusion

Table 1. Some diffusion parameters of hydrogel systems.

Peleg proposed a two-parameter model to describe water absorption by hydrogels [26]:

$$Q_t = Q_0 \pm \frac{t}{k_1 + k_2 t'} \tag{4}$$

where Q_0 is the swelling content at $Q = 0$ (g/g d.b.), Q_t is the swelling content at any time (g/g d.b.), t is the swelling time (s), k_1 is the kinetic constant of the model (h(g d.b.)/g), and k_2 is a characteristic constant of the model (g d.b.)/g). In Eq. (4), "±" becomes "+" if the process is absorption or adsorption and "−" if the process is drying or desorption.

This section examines the possibility that diffusion-controlled swelling follows fist-order kinetics, as is frequently assumed. According to first-order kinetics, the rate of swelling at any given time (t) is directly proportional to the uptake of swelling medium that has yet to occur before the maximum or equilibrium uptake (Q_{eq}) has been reached. If Q is the uptake at time t, $Q_{eq} - Q$ is the unrealized uptake of swelling medium. If k is the proportionality constant between the rate of swelling and the unrealized swelling capacity, then [27]:

$$\frac{dQ}{dt} = k_1 \left(Q_{eq} - Q_t\right), \tag{5}$$

which integrates to:

$$ln\left(\frac{Q_{eq}}{Q_{eq} - Q_t}\right) = k_1 t. \tag{6}$$

The Schott's second-order equation for swelling is [27]:

$$\frac{dQ}{dt} = k_2 \left(Q_{eq} - Q_t\right)^2, \tag{7}$$

where Q_{eq} is the equilibrium water swelling ratio, Q_t is the water swelling ratio at time t, k_2 is the swelling rate constant, respectively. After definite integration between the limits $Q = 0$ at t = 0 and Q = Q at t = t and rearrangement, Eq. (7) can be rewritten as follows:

$$\frac{t}{Q_t} = \frac{1}{Q_{eq}^2 k_2} + \frac{1}{Q_{eq}} t. \tag{8}$$

To test the all kinetics models described above, $ln\left(\frac{Q_{eq}}{Q_{eq} - Q}\right)$ vs t, t/Q vs t, and 1/Q vs 1/t graphs were plotted for analyzed hydrogel composites. Values of swelling rate constant and equilibrium swelling ratio, which were calculated from the slope and the intersection of the lines, respectively, are presented in **Table 2**.

Figure 5 shows the comparison graphs of experimental data on the swelling kinetics in water with the three models described above. Many authors believe that the swelling kinetics behavior of ionic hydrogels (anionic and cationic hydrogels) is in good accordance with Schott's second-order diffusion kinetics [28, 29]. As can be seen from the **Table 2**, the most accurate process of swelling of hydrogel filled with bentonite describes the first-order kinetic model. Values of the calculated equilibrium degree of swelling $Q_{eq\,calc}$ are in good agreement with the experimental data. A second-order kinetic model also yields adequate results. For gels with pectin, all models give an acceptable result, although the most accurately nevertheless the kinetics of swelling describes a model of the first order.

	Content of bentonite in hydrogel composites (%)					Content of pectin in hydrogel composites (%)			
	0	1	2	3	4	1	2	3	4
Peleg's model									
k_1	2.1810	1.3559	1.4929	1.4593	1.7098	1.1032	0.9791	1.6753	1.7822
k_2	0.004	0.0069	0.0071	0.0036	0.0080	0.0044	0.0042	0.0044	0.0056
R^2	0.9989	0.9977	0.9973	0.9995	0.9938	0.9987	0.9988	0.9984	0.9994
Q_{exp}	247.0146	259.77287	278.517	254.182	237.26	185.447	200.764	161.373	158.642
Q_{calc}	276.749	140.4422	136.1533	259.1549	120.0033	216.941	226.848	210.621	170.177
First-order absorption kinetic model									
k_{R1}	0.0008	0.0012	0.0011	0.0020	0.0017	0.0047	0.0040	0.0042	0.0027
R^2	0.9808	0.9968	0.9957	0.9814	0.9920	0.9972	0.9995	0.9967	0.9995
Q_{calc}	244.305	259.474	277.954	254.178	237.243	185.447	200.764	161.373	158.642
Second-order absorption kinetic model									
k_{R2}	1.03E−5	0.93E−5	0.76E−5	1.23E−5	1.02E−5	3.89E−5	3.48E−5	3.50E−5	2.63E−5
R^2	0.9975	0.9960	0.9955	0.9961	0.9948	0.9995	0.9996	0.9992	0.9994
Q_{calc}	261.096	280.899	305.810	278.551	265.251	189.753	207.039	166.389	166.666

Table 2. Swelling kinetic parameters for poly(AAm-co-AAc)/bentonite and poly(AAm-co-AAc)/pectin hydrogel composites in distilled water.

http://dx.doi.org/10.5772/intechopen.71190

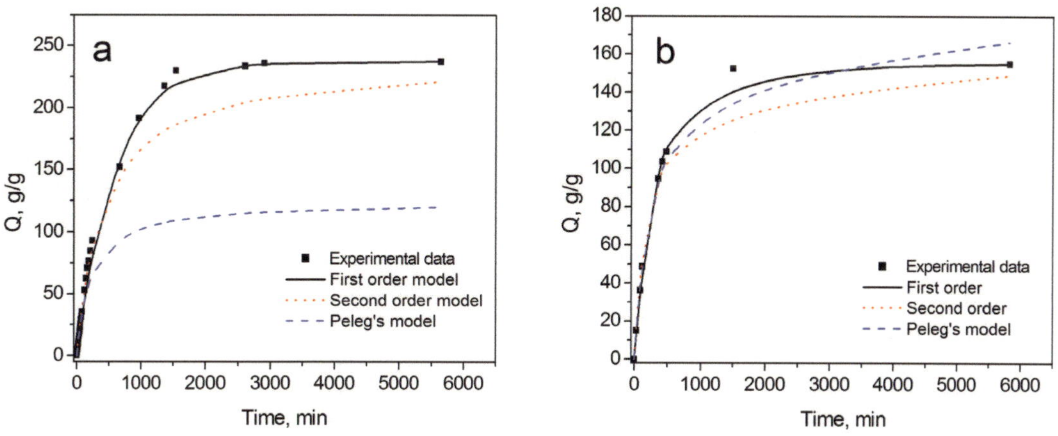

Figure 5. Comparison of the experimental swelling data with first-order, second-order kinetic models, and Peleg's model for a poly(AAm-co-AAc)/bentonite 4% (a) and poly(AAm-co-AAc)/pectin 4% (b).

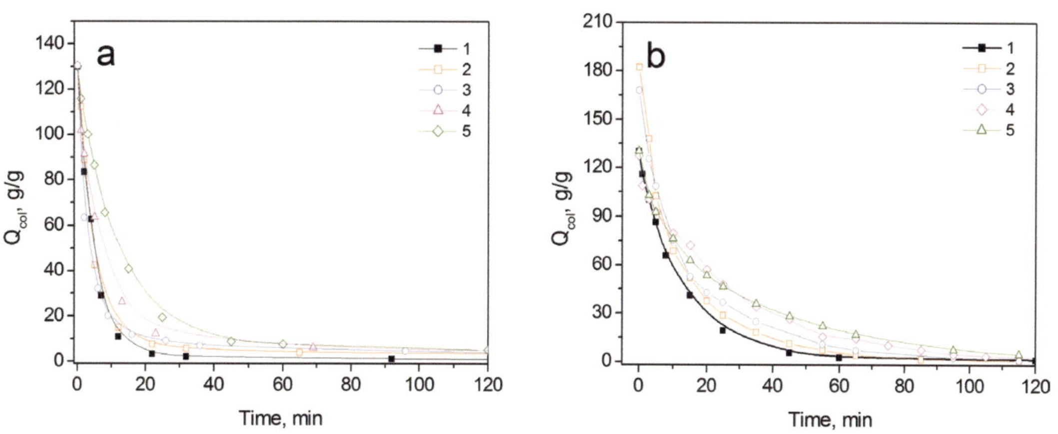

Figure 6. Collapse curves of poly(AAm-co-AAc)/bentonite composites and poly(AAm-co-AAc)/pectin composites filled with 0% (1), 1% (2), 2% (3), 3% (4), and 4% (5) bentonite (a) and pectin (b) in a 1 M solution of calcium chloride.

3.2.4. Collapse kinetics

Deswelling (or collapse) kinetic experiments were carried out by immersing a known amount of the fully swollen hydrogels at 25°C of the distilled water in various solutions with different ionic strength. At predetermined time intervals, the hydrogels were taken out and weighted. The percentage water retention was calculated using the following equation:

$$Q = \frac{m_2 - m_0}{m_0}, \tag{9}$$

where m_2 (g) and m_0 (g) are the weights of swollen hydrogel and of the original dry hydrogel in solution with different ionic strength.

Figure 6 shows the kinetic curves of collapse in a 1 M solution of calcium chloride. It can be seen from the figure that the collapse of composite gels occurs with both to bentonite, and to pectin. However, it can be noted that the presence of bentonite and pectin in polyelectrolyte hydrogel composites to some extent prevents collapse, reducing the rate of desorption of water into the solution. This effect is most typical for polymer hydrogels with higher filler content. **Figure 6** shows that the collapse curves have a high collapse rate at the initial section for an unfilled hydrogel than for a polymer composite with 5 wt.% bentonite and 4% pectin. A plausible explanation is that the spatial interactions between the bentonite plates exclude the further collapse of the gels. As can be seen from **Figure 6b**, a hydrogel with a semi-interpenetrating network has a lower collapse rate, which indicates large spatial interactions between the polymer network and the polysaccharide chains, as compared to bentonite particles.

The first-order collapse rate constants calculated from Eq. (6) are shown in **Table 3**. It can be seen from the table that the rate constants of collapse in various salts decrease with increasing percentage of bentonite and pectin filling in the polymer composite, which indicates a decrease the rate of collapse depending on the content of fillers.

Sample	**Percentage of filler in the composite (%)**					
		0	**1**	**2**	**3**	**4**
Peleg's model						
Poly(AAm-co-AAc)/bentonite	k_1	0.06488	0.04719	0.02848	0.02904	0.01437
	k_2	0.00764	0.00569	0.00706	0.00795	0.0077
	R^2	0.99596	0.98497	0.98842	0.99	0.99871
Poly(AAm-co-AAc)/pectin	k_1		0.02965	0.03337	0.08836	0.04875
	k_2		0.00274	0.00308	0.00806	0.01068
	R^2		0.98704	0.99775	0.99516	0.9369
First-order kinetic model						
Poly(AAm-co-AAc)/bentonite	k_1	0.029	0.04313	0.02999	0.03526	0.05486
	R^2	0.86309	0.86428	0.88308	0.8132	0.97665
	$\mathbf{k_1^*}$	**0.08591**	**0.12385**	**0.14159**	**0.11425**	**0.07411**
	$\mathbf{R^{2*}}$	**0.92486**	**0.97008**	**0.97526**	**0.99654**	**0.99946**
Poly(AAm-co-AAc)/pectin	k_1		0.06199	0.04873	0.03695	0.03544
	R^2		0.99515	0.98231	0.99362	0.99761
	$\mathbf{k_1^*}$		**0.11433**	**0.08432**	**0.04918**	**0.0348**
	$\mathbf{R^{2*}}$		**0.99147**	**0.99804**	**0.98799**	**0.96057**

*Kinetic constants are calculated at the initial stage of the collapse of hydrogels (0–50 min).

Table 3. The rate constants of the collapse, calculated from the Peleg's model and the first-order kinetic model.

Figure 7 compares the experimentally obtained dependence of Q on the time of the collapse process with the theoretical dependences obtained in the approximation of the experimental data to the dependences (6) and (4) corresponding to the kinetic models of the pseudo-first-order collapse and Peleg, respectively. As can be seen from the graph, the kinetics of the collapse of the poly(AAm-co-AAc)/bentonite hydrogel in salt solutions is most appropriately described by Peleg's model (**Figure 7a**). However, to describe the kinetics of the hydrogel collapse with a semi-interpenetrating network—poly(AAm-co-AAc)/pectin, the pseudo-first-order model is most suitable.

3.2.5. *Equilibrium sorption studies*

It is known that bentonite possesses good sorption properties and is used to purify water from pollutants. Articles [30, 31] cite data that bentonite is added to composite materials in order to increase the sorption properties of the material. It is also known that pectin is an enterosorbent [32]. Pectins have complex-forming ability based on the interaction of a molecule of pectin with ions of heavy metals and radionuclides. Due to the presence of a large number of free carboxyl groups in molecules, it is the low-esterified pectins that are most effective.

In the present work, the absorption properties of the synthesized polymer compositions in aqueous salt solutions of $Pb(NO_3)_2$ were to be studied. To observe the sorption of Pb^{2+} ions onto poly(AAm-co-AAc) hydrogel systems containing bentonite and pectin, the hydrogel systems were placed in aqueous solutions of $Pb(NO_3)_2$ and allowed to equilibrate for 1 day at 25°C. The concentration of sorbed ions was determined by the voltammetric method.

Equilibrium adsorption isotherms of poly(AAm-co-AAm), poly(AAm-co-AAc)/bentonite and poly(AAm-co-AAc)/pectin hydrogel systems are presented in **Figure 8**.

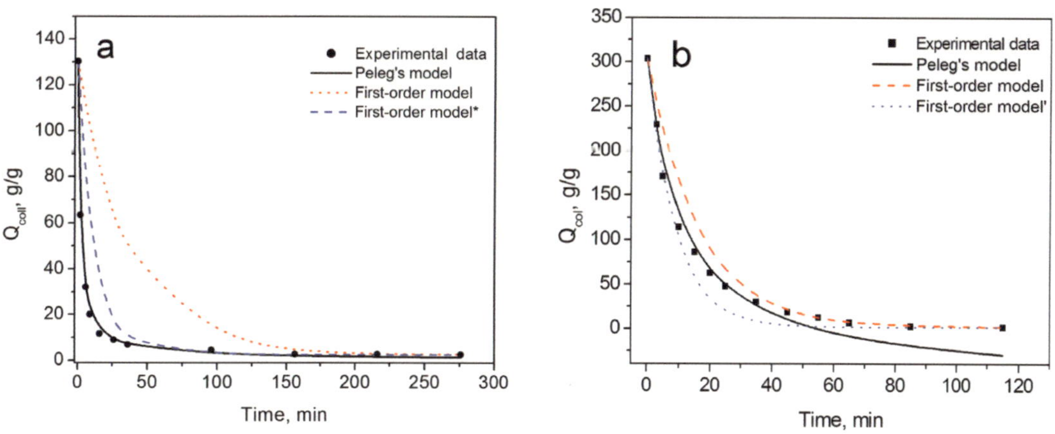

Figure 7. Experimental data and data obtained with first-order models and Peleg's model for poly(AAm-co-AAc)/bentonite 1% (a) and poly(AAm-co-AAc)/pectin 1% (b) hydrogel composites.

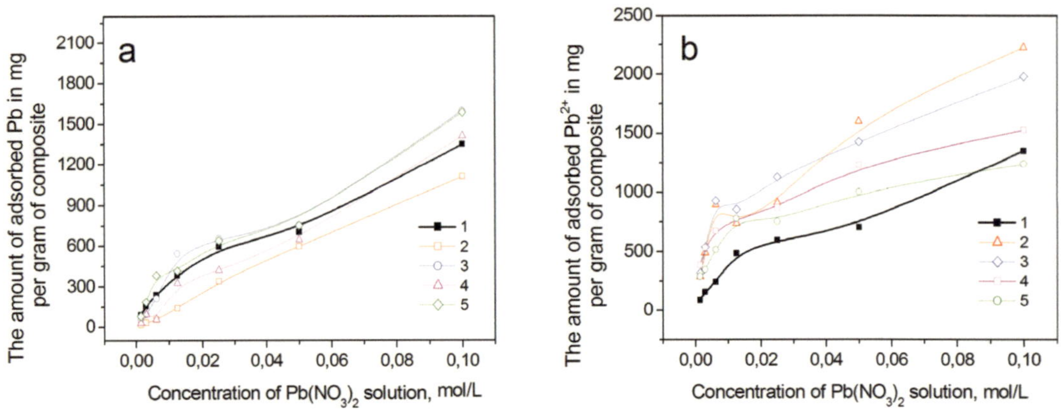

Figure 8. Sorption of lead ions from solutions of various concentrations of $Pb(NO_3)_2$ poly(AAm-co-AAc) hydrogel (1) and poly(AAm-co-AAc)/pectin (a) hydrogels with 1% (2), 2% (3), 3% (4), and 4% (5) pectin and poly(AAm-co-AAc)/ bentonite (b) hydrogels with 1% (2), 2% (3), 3% (4), and 4% (5) bentonite.

In **Figure 8**, the heavy metal ions removal capacity (mg amount of sorption Pb^{2+} per unit mass) of the hydrogel systems is increased with the increasing concentration $Pb(NO_3)_2$ sorbed onto unit dry mass of the gel. Adsorption of heavy metal ions occurs due to ionic and coordination interactions with charged hydrogel groups (chemisorption).

At first, complexation takes place mainly on the surface of the hydrogel, as evidenced by the high initial rate of sorption of metal ions, possible to observe inhomogeneous distribution of ionic groups, which leads to uneven swelling of the hydrogel by volume. The experimental results show that the hydrogel-filled composites filled with pectin have greater sorption ability than the unfilled hydrogel (**Figure 8a**). The increase in the sorption of lead ions by pectin-containing hydrogels is via additional complex formation due to the presence of a large number of free carboxyl groups in the pectin molecules. Thus, from a solution with a concentration of 0.1 M lead nitrate, composites with 1 and 3% pectin are adsorbed by 20% of lead ions more than the uncharged hydrogel. However, as **Figure 8b** shows, hydrogels containing bentonite absorb a larger amount of lead ions. Moreover, the sorption capacity of bentonite-containing hydrogels is 25% higher than that of polysaccharide-containing hydrogels (as moisture absorbing properties).

4. Conclusion

Nowadays, there is a trend in development of multifunctional nanocomposites based on different types of hydrogels acting as a matrix for various nanomaterials. Crosslinked hydrophilic polymers are capable of absorbing large volumes of waters and salt solutions. Therefore, most of the modern work is devoted to the development of new composite materials based on hydrogels with improved sorption properties.

In the chapter of the book, two representatives of hydrogel composites are considered—a composite with dispersed filler and a composite with a semi-interpenetrating network and their sorption properties are compared.

Thus, incorporation of hydrophilic group containing chemicals, such as acrylic acid, polymers such as pectin, and clay such as bentonite in AAm hydrogels, can be obtained successively by the free radical solution polymerization method. Multifunctional crosslinkers such as MBA were used in the polymerization process. Poly(AAm-co-AAc)/bentonite and poly(AAm-co-AAc)/pectin hydrogel systems have showed high water absorbency. Some swelling and diffusion properties were discussed for different semi-IPNs and hydrogels prepared under various formulations. To determine the sorption characteristics of heavy metal ions such as Pb into the hydrogel systems, some sorption parameters have been calculated.

Acknowledgements

The authors acknowledge Dr. Ivanova A.I., Tver State University, Research Laboratory of Electron Microscopy.

Author details

Mayya V. Uspenskaya, Vera E. Sitnikova, Michael A. Dovbeta, Roman O. Olekhnovich and Igor Yu. Denisyuk*

*Address all correspondence to: denisiuk@mail.ifmo.ru

Department of IT in Fuel and Energy Industry, ITMO University, St. Petersburg, Russia

References

[1] Zhou G, Luo J, Liu C, Chu L, Ma J, Tang Y, Zeng Z, Luo S. A highly efficient polyampholyte hydrogel sorbent based fixed bed process for heavy metal removal in actual industrial effluent. Water Research. 2016;**89**:151-160. DOI: 10.1016/j.watres.2015.11.053

[2] Hou Y, Brower M, Pollard D, Kanani D, Jacquemart R, Kachuik B, Stout J. Advective hydrogel membrane chromatography for monoclonal antibody purification in bioprocessing. Biotechnology Progress. 2015;**31**(4):974-982. DOI: 10.1002/btpr.2113

[3] Park Y, Lee KH. Preparation of water-swollen hydrogel membranes for gas separation. Journal of Applied Polymer Science. 2001;**18**(10):1785-1791. DOI: 10.1002/app.1274

[4] Nakagawa K, Nishimoto N. Cryotropic gel formation for food nutrients encapsulation—A controllable processing of hydrogel by freezing. Procedia Food Science. 2011;**1**:1968-1972. DOI: 10.1016/j.profoo.2011.09.289

[5] Hoare TR, Kohane DS. Hydrogels in drug delivery: Progress and challenges. Polymer. 2008;**49**:1993-2007. DOI: 10.1016/j.polymer.2008.01.027

[6] Caló E, Khutoryanskiy VV. Biomedical applications of hydrogels: A review of patents and commercial products. European Polymer Journal. 2015;**65**:252-267. DOI: 10.1016/j.eurpolymj.2014.11.024

[7] Pavlyuchenko VN, Ivanchev SS. Composite polymer hydrogels. Polymer Science, Series A. 2009;**51**(7):743-760. DOI: 10.1134/S0965545X09070013

[8] Flory PJ. Principles of Polymer Chemistry. Ithaca, NY: Cornell University Press; 1953. p. 688

[9] Flory PJ, Rehner J. Statistical mechanics of cross-linked polymer networks II. Swelling. The Journal of Chemical Physics 1943;**11**(11):521-526. DOI: http://dx.doi.org/10.1063/1.1723792

[10] Katchalsky A, Lifson S, Eisenberg H. Equation of swelling forpolyelectrolyte gels. Journal of Polymer Science. 1951;**7**(5):571-574. DOI: 10.1002/pol.1951.120070513

[11] Katchalsky A, Michaely I. Polyelectrolyte gels in salt solutions. Journal of Polymer Science. 1955;**15**(79):69-86

[12] Peppas NA, Huang Y, Torres-Lugo M, Ward JH, Zhang J. Physicochemical foundations and structural design of hydrogels in medicine and biology. Annual Review of Biomedical Engineering. 2000;**2**:9-29. DOI: 1523-9829/00/0825-0009$14.00

[13] Everaer R, Kremer K. Elastic properties of polymer networks. Journal of Molecular Modeling 1996;**2**(9):293-299. DOI: http://dx.doi.org/10.1007/s0089460020293

[14] Grosberg AY, Khokhlov AR, Pande VS. Statistical Physics of Macromolecules. 1st ed. New York: AIP Press; 1994. p. 350

[15] Muniz EC, Geuskens G. Compressive elastic modulus of polyacrylamide hydrogels and semi-IPNs with poly(N-isopropylacrylamide). Macromolecules. 2001;**34**(13):4480-4484. DOI: 10.1021/ma001192l

[16] Horkay F, Hecht A-M, Geissler E. Effect of cross-links on the swelling equation of state: Polyacrylamide hydrogels. Macromolecules. 1989;**22**(4):2007-2009. DOI: 10.1021/ma00194a081

[17] Cohen Y, Ramon O, Kopelman IJ, Mizrahi S. Characterization of inhomogeneous polyacrylamide hydrogels. Journal of Polymer Science Part B: Polymer Physics. 1992;**30**(9):1055-1067. DOI: 10.1002/polb.1992.090300913

[18] Horkay F, Hecht A-M, Zrinyi M, Geissler E. Effect of cross-links on the structure of polymer gels. Polymer Gels and Networks. 1996;**4**(5-6):451-465. DOI: https://doi.org/10.1016/S0966-7822(96)00041-X

[19] Lowman AM, Peppas NA. Hydrogels. In: Mathiowitz E, editor. Encyclopedia of Controlled Drug Delivery. New York: Wiley; 1999. pp. 397-418

[20] Berger J, Reist M, Mayer JM, Felt O, Peppas NA, Gurny R. Structure and interactions in covalently and ionically crosslinked chitosan hydrogels for biomedical applications. European Journal of Pharmaceutics and Biopharmaceutics. 2004;**57**(1):19-34. DOI: https://doi.org/10.1016/S0939-6411(03)00161-9

[21] Magalhaes ASG, Neto MPA, Bezerra MN, Ricardo NMPS, Feitosa JPA. Application of FTIR in the determination of acrylate content in poly(sodium acrylate-*co*-acrylamide) superabsorbent hydrogels. Quimica Nova. 2012;**35**(7):1464-1467. DOI: http://dx.doi.org/10.1590/S0100-40422012000700030

[22] Murugan R, Mohan S, Bigotto A. FTIR and polarized Raman spectra of acrylamide and polyacrylamide. Journal of the Korean Physical Society. 1998;**32**(4):505-512

[23] Peppas NA, Franson NM. The swelling interface number as a criterion for prediction of diffusional solute release mechanisms in swellable polymers. Journal of Polymer Science, Polymer Physics Edition. 1983;**21**(6):983-997. DOI: 10.1002/pol.1983.180210614

[24] Ritger PL, Peppas NA. A simple equation for description of solute release I. Fickian and non-Fickian release from non-swellable devices in the form of slabs, spheres, cylinders or discs. Journal of Controlled Release. 1987;**5**(1):23-36. DOI: https://doi.org/10.1016/0168-3659(87)90034-4

[25] Mohammad Karimi . Diffusion in Polymer Solids and Solutions. In: Jozef MarkoÅi (Ed.), Mass Transfer in Chemical Engineering Processes. InTech; 2011. DOI: 10.5772/23436. Available from: https://www.intechopen.com/books/mass-transfer-in-chemical-engineering-processes/diffusion-in-polymer-solids-and-solutions

[26] Peleg M. An empirical model for the description of moisture sorption curves. Journal of Food Science. 1988;**53**(4):1216-1217. DOI: 10.1111/j.1365-2621.1988.tb13565.x

[27] Schott H. Kinetics of swelling of polymers and their gels. Journal of Pharmaceutical Sciences 1992;**82**(5):467-470. DOI: http://dx.doi.org/10.1002/jps.2600810516

[28] Bamgbose JT, Bamigbade AA, Adewuyi S, Dare EO, Lasisi AA, Njah AN. Equilibrium swelling and kinetic studies of highly swollen chitosan film. Journal of Chemistry and Chemical Engineering. 2012;**6**:272-283

[29] Trivedi JH. Synthesis, characterization and swelling behavior of superabsorbent hydrogel from sodium salt of partially carboxymethylated tamarind kernel powder-g-PAN. Journal of Applied Polymer Science. 2013;**129**(4):1992-2003. DOI: 10.1002/app.38910

[30] Kaplan M, Kasgoz H. Hydrogels nanocomposite sorbents for removal of basic dyes. Polymer Bulletin. 2011;**67**(7):1153-1168. DOI: https://doi.org/10.1007/s00289-011-0444-9

[31] Zhou SH, Yang JG, Wu CP. Synthesis and swelling properties of poly (N,N′-diethylacrylamide)-clay nanocomposites. Acta Polymerica Sinica. 2003;**3**:326-329

[32] Trachtenberg IM. Pectin-containing oral adsorbents against radionuclides and heavy metals action. Physycian Matter. 1992;**4**:23-29

Section 3

Mechanical Properties of Polymerization

Chapter 6

Thermoplastic Foams: Processing, Manufacturing, and Characterization

Mihrigul Altan

Additional information is available at the end of the chapter

http://dx.doi.org/10.5772/intechopen.71083

Abstract

Polymer foams have wide application area due to their light weight, resistance to impact, high thermal insulation, and damping properties. Automotive, packing industry, electronic, aerospace, building construction, bedding, and medical applications are some of the fields that polymer foams have been used. However, depending on their cell structure—open or closed cell—polymer foams have different properties and different application areas. In this work, the most used thermoplastic foams with closed cells such as polypropylene, polyethylene, and polystyrene or polylactic acid have been focused. Their melt strength, degree of crystallinity for semi-crystalline ones, and viscosity have great importance on cell morphology. Cells in small diameter with high dense in polymer matrix are preferable. However, obtaining fine cells is not easy in each case, and it is still under investigation for some polymers. There are several ways to improve cell morphology, and one of them is addition of nanoparticle to the polymer. During foaming process, nanoparticles behave like nucleating agent that cells nucleate at the boundary between polymer and the nanoparticle. Besides, foaming agents contribute the homogenous dispersion of the nanoparticles in the polymer matrix, and this improves the properties of the polymer foams and generates multifunctional material as polymer nanocomposite foams.

Keywords: thermoplastic foams, closed cell foams, foam processing, nanofiller, cell morphology, polymer foam nanocomposites

1. Introduction

Polymeric foams are widely used in different fields due to their lightness, reduced thermal conductivity, high-energy absorption, and excellent strength/weight ratio. Application area of the polymer foams has high variety such as transportation, bedding, carpet underlay, textile, toys, sport instruments, insulation appliances, and construction, biomedical, and automotive

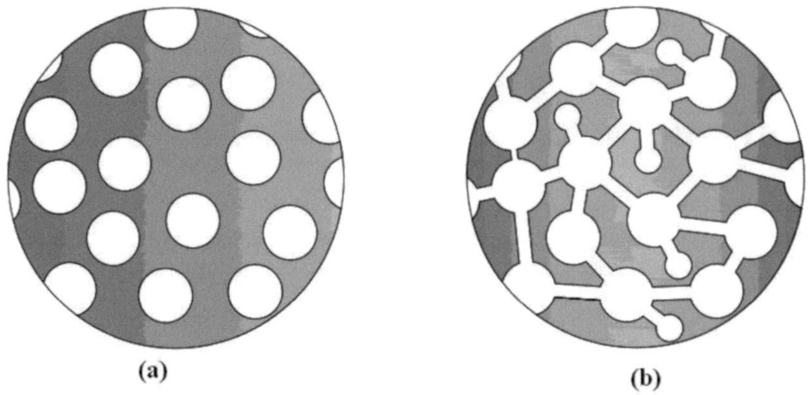

Figure 1. Illustration of polymer foam cellular structures (a) closed cell type (b) open cell type.

sectors [1–5]. A polymer foam is basically a polymer-and-gas mixture, which gives the material a microcellular structure. Polymer foams can be flexible or rigid due to their cell geometry such as open cells or closed cells (**Figure 1**). If the gas pores roughly spherical and separated from each other by the polymer matrix, then this type is called closed cell structure. On the contrary, if the pores are interconnected to each other to some extent which provides passage of fluid through the foam, then this is called open cell structure. A close cell structure is a good candidate to be a life jacket material, while an open cell structure would be waterlogged. The open cell foams are for bedding, acoustical insulation car seating, and furniture, while the closed cell foams are suitable for thermal insulation, and they are generally rigid, which makes them a preferable lightweight material for automotive and aerospace [6–9].

The development of polymeric foams started with the macrocellular polystyrene foams having cell size above 100 μm in the 1930s [10]. Developments continued for providing finer cells, and solid-state batch foaming method was applied and foam cells less than 100 μ in diameter were obtained in the 1980s. Since then, polymer foam processing and shaping methods have developed speedily. Besides polystyrene foams, polyurethane has become popular. However, in this work, the most used thermoplastic foams with closed cell structure are focused. The cell generation, cell size and density, mechanical properties, and shaping processes of thermoplastic foams are given in detail. The effect of nanoparticle addition is also discussed in generation of multifunctional materials, polymer nanocomposite foams.

2. Thermoplastic foam processing methods

The principle of foaming processes includes the steps of polymer saturation or impregnation with a foaming agent, providing super saturated polymer-gas mixture by either sudden increment of temperature or decrease in pressure, cell growth, and stabilization [11]. In thermoplastic foaming processes, it is important to obtain foams with closed cell structure with thin polymer cell walls covering each cell. In order to provide this structure, cell growth must be controlled through the process. Temperature limit is critical in obtaining microcellular structure.

If temperature is higher excessively, then melt strength of the polymer can be low-inducing cell rupture. On the other hand, if temperature is too low, this will result in longer foaming times and increment in viscosity of the polymer. As a consequence, cell growth will be restrained, and insufficiently foamed products will be obtained. Therefore, the process conditions have serious importance on cell morphology of the polymer foams. The most known thermoplastic foaming processes are batch foaming, extrusion foaming, and foam-injection molding.

2.1. Batch foaming

Batch foaming can be applied in two different methods as follows, pressure-induced method and temperature-induced method. In pressure-induced method (**Figure 2**), polymer is saturated with blowing agent in an autoclave, and then, cell nucleation is done by sudden depressurization of the system to atmospheric pressure. The final cell morphology is obtained by either cooling the polymer in a solvent or by cooling it within air [10].

In temperature-induced batch foaming (**Figure 3**), the beginning of the process is similar to pressure-induced foaming but at lower temperatures. After saturation is completed, the sample is taken out of the autoclave and put into hot oil bath between the temperatures of 80–150°C for a period of time in order to obtain cell generation. After this step, the sample is put into a cooling bath of water or a solvent. The important point in batch foaming is the geometry of the plastic samples. They are generally a circular disc, rectangular, or square shape with the thickness between 0.5–3 mm not to hinder gas diffusivity [10].

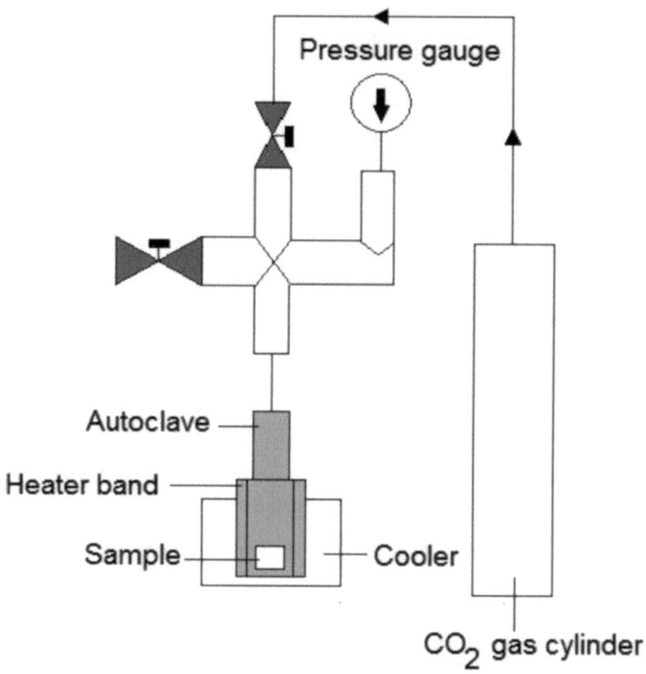

Figure 2. Pressure-induced batch foaming.

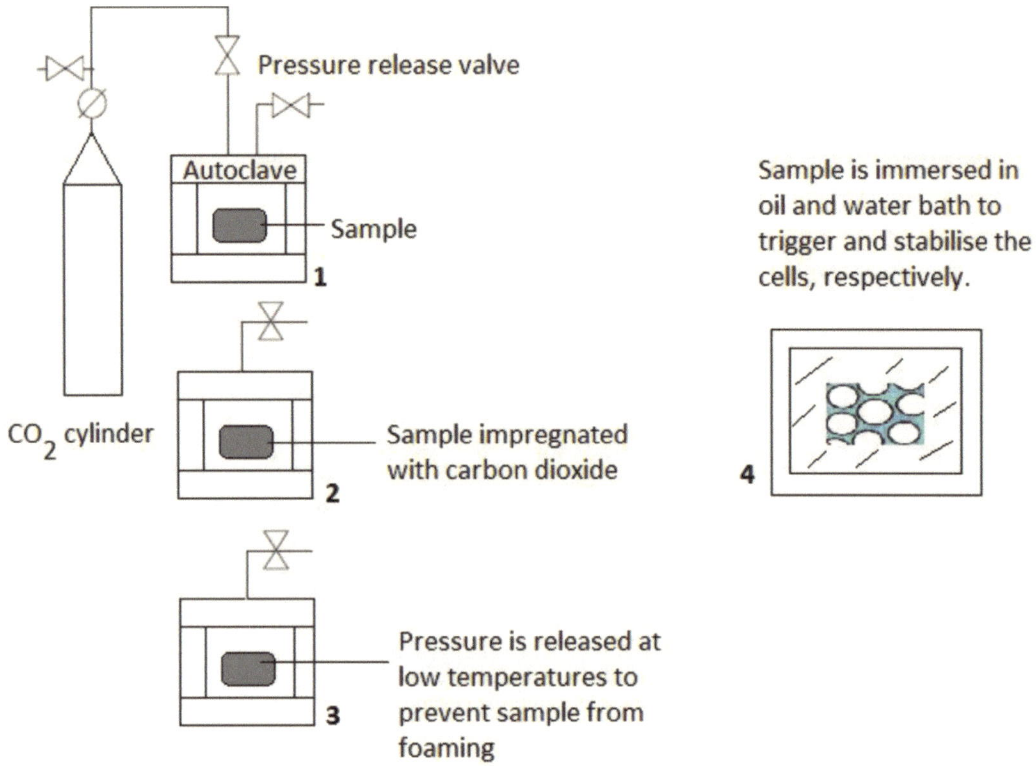

Figure 3. Steps of temperature-induced batch foaming.

2.2. Foam extrusion

In foam extrusion, a tandem line extrusion machine is equipped with a gas supply as shown in **Figure 4**. Typical product types are thermoplastic-based foamed sheets, pipes, and expanded tubes. The pellets supplied from the hopper into the barrel are melted under high pressure and blowing agent. CO_2 gas in supercritical condition is injected into the polymer. Due to the high pressure in the barrel, nucleation of the foam cells is prevented. As the polymer exists from the die, foam cells are generated by the sudden pressure drop. The final step is cooling, calibration, and cutting of the extruded foams [11, 12].

The extrusion foaming process can be either physical or chemical foaming. In **Figure 4**, physical foaming is shown that a gas supply is integrated to the extruder. In industrial applications, chemical foam extrusion is also applied due to its cheapness in tooling. In chemical foam extrusion, polymer pellets and chemical foaming agent are mixed through the barrel, and the heat in the barrel decomposes the chemical foaming agent resulting in gas which provides expansion of the polymers as it exits the die. Melt temperature is critical in decomposition of the foaming agent. The pressure must be high enough in order to keep the dissolved gas in the polymer before it exits the die. If the pressure and temperature are not set correctly, foaming agent will not be decomposed and can induce left particles or agglomerations of foaming agent, which can lead to poor cell morphology and poor surface quality [13]. The most known chemical

http://dx.doi.org/10.5772/intechopen.71083

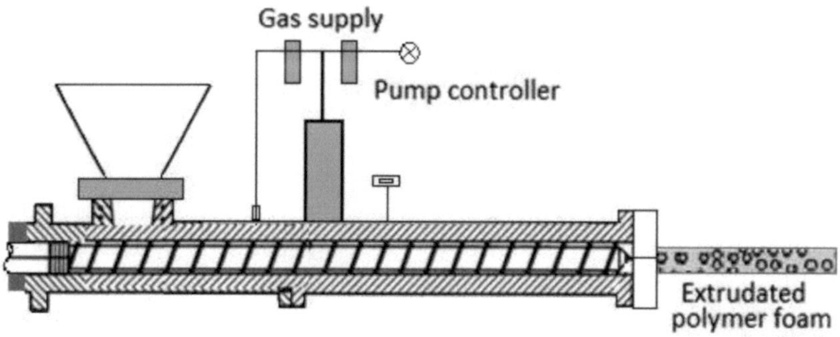

Figure 4. Foam extrusion.

foaming agent is azodicarbonamide (ADC), an exothermic chemical foaming agent. It releases high-amount of N_2 gas together with CO_2 in lower amount into the polymer. However, due to the toxic byproducts of ACD, endothermic type commercial foaming agents are being used, such as Clariant's Hydrocerol [13, 14].

2.3. Foam injection molding

Foam injection molding is similar to conventional injection molding, but an additional gas unit is integrated to the injection molding machine if physical foaming is applied (**Figure 5**). There are currently three widely known foam injection–molding technologies available to produce microcellular foams using CO_2 as a physical blowing agent. They are MuCell by Trexel Inc. (USA), Optifoam by Sulzer Chemtech AG (Switzerland), and ErgoCell by Demag (Germany) [15, 16].

Foam injection molding has some critical points to be considered. One of them is the presence of the back pressure. If back pressure is not applied, polymer-gas mixture would move the screw axially and instability in dosing of the polymer would be seen. Also, foaming agent would expand in the plasticization unit and leak out during injection. This would prevent cell

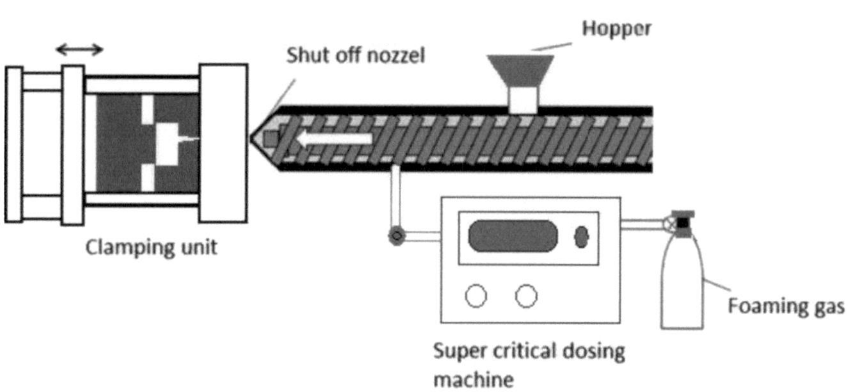

Figure 5. Foam injection molding.

generation in the polymer. The second critical point in foam injection molding is the selection of needle shut off nozzle that prevents the leak out of the nozzle and gas loss [16].

In foam injection molding, physical and chemical foaming can be applied. In chemical foaming, chemical foaming agent is added in solid form either from the hopper of the injection molding machine with the polymer pellets or during plasticization of the polymer through the barrel. Foaming agent dissolve through the process. Physical foaming agents are injected directly into the molten polymer. The difference in comparison to foam extrusion is the motion of the screw. In foam extrusion, the screw rotation pushes the melt forward and then out of the extruder die, but in foam injection molding, screw rotates and moves backward due to the collection of a pool of gas-polymer mixture at the tip of the screw. Then, polymer-gas mixture is injected into the cavity under. In physical foaming, the high pressure and high temperature in plasticization unit provide supercritical state of the foaming agent [17]. Gases like nitrogen (N_2) and carbon dioxide (CO_2) are used as physical foaming agents, and they are applied in an overcritical state in order to obtain high degree of solubility in the molten polymer [17]. In supercritical fluid state, fluid has low viscosity, low surface tension, and high diffusion properties that all these provide excellent solubility in the polymer. Depending on this, improved cell morphology is achieved. Carbon dioxide has supercritical point of 73.84 bar with 37°C, and nitrogen has 33.90 bar with −147°C. In **Figure 6**, the supercritical phase of carbon dioxide is shown.

In order to control the dosing of gas, supercritical dosing machine is integrated to the system as shown in **Figure 5**. Furthermore, high back pressure is necessary during plasticization for dosing and homogenizing the foaming agent in the polymer melt [17]. For these reasons, a specially equipped machine similar to the conventional injection molding is necessary in foam injection molding, as shown in **Figure 5**.

The high equipped and expensive systems in physical foaming of polymer foams processing is costly. On the other hand, chemical foaming is less complicated and can liberate gases under certain processing conditions either due to chemical reaction or thermal

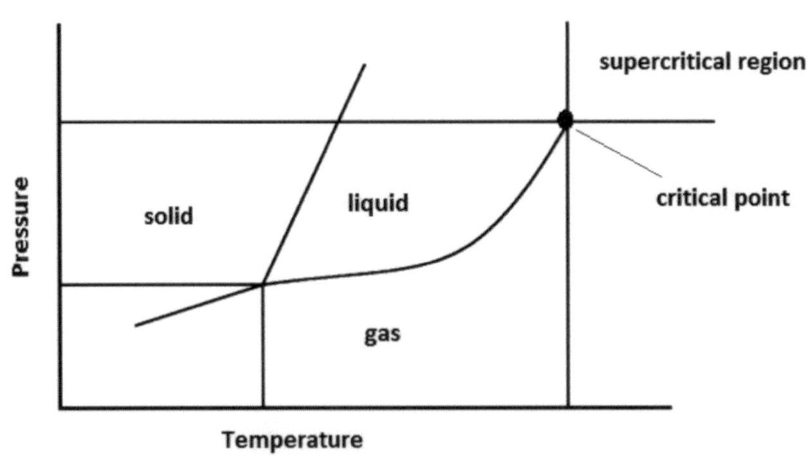

Figure 6. Supercritical fluid CO_2.

decomposition [13]. Chemical foaming agents are added to polymer either prior to or during plasticization, similar to foam extrusion by chemical foaming agents. They can exothermic or endothermic. Exothermic types release energy during a reaction which is dissipated through the plasticization unit. As the activation temperature is reached, no energy is needed to be added, and reaction continues until foaming agent finishes its reacting completely. In the usage of endothermic foaming agents, energy must be continuously applied in the form of heat, so that the reaction does not stop. Azodicarbonamide (AC) is the most known exothermic foaming agent high gas yield. It has decomposition temperatures between 170 and 200°C [13]. Sodium bicarbonate and zinc bicarbonate are the most common endothermic blowing agents. In last few years, a commercial foaming agent, Hydrocerol, has been extensively employed as endothermic foaming agent. Hydrocerol has decomposition temperatures between 160 and 210°C and can be added directly into the hopper of an injection-molding machine in the form of pellets in proportions from 1% to 4 wt.% [13].

As a summary, the comparison of three foaming processes is given in **Table 1**.

2.3.1. *Morphology of foams*

In batch foaming, uniform cell size and homogenous distribution of the cells can be gained. The polymer parts are foamed in solid state in batch foaming; therefore, a very tiny skin layer is formed as unfoamed. In foam extrusion, uniform cell size is possible to obtain, but cells in the core of the extrudated part may be larger due to the instability in cooling stage. On the other hand, morphology of the cells in foam injection molding shows locally different character through the thickness of the molded part due to the variation in temperature from mold wall to the core of the part. The mold wall has low temperature than that of polymer melt leading to a sudden freezing of the polymer close to the mold wall. In this zone, which is called skin layer, cell generation is inhibited. The foaming agent dissolved in the polymer remains in the skin layer and diffuses out of the polymer. As a result, a frontal flow in the core of the polymer melt is generated as shown in **Figure 7**. This results in a compact skin layer and a foamed core [18–20].

The morphology of the polymer foams is important and affects the mechanical strength of the material. Cells in small in diameter enhance the mechanical strength when compared with the larger cells. The density of the foams can be defined by the distance between neighboring cells. It is generally defined as 0.04–0.30 g/cm^3. Cell distances have shown that they have a definite influence on the mechanical properties of thermoplastic foams [16, 19]. The morphology of the foam injected part can be divided into five zones. As shown in **Figure 8**, the zones from one mold wall to the other mold in the cavity are skin layer-outer foam core-inner foam core-outer foam core-skin layer. Inner core has the cells with the largest diameter due to the slow cooling rate of the material in the core region, and cells have time to expand [16, 18–20].

Shortly, morphology of the foam injection moldings has great importance on the properties of the polymer foams such as mechanical, optical, and thermal conductivity. For this reason, setting of the injection molding parameters correctly and dosing of foaming agent preciously

Criteria	Batch foaming	Foam extrusion	Foam injection molding
Amount of material needed	Small amount (in g)	Larger amount (in kg)	Larger amount (in kg)
Pre-molding	Necessary	No needed, molding tool is in the process itself	No needed, molding tool is in the process itself
Sample state during gas loading/saturation temperature	Solid	Melt state	Melt state
Cell density range (cells/cm³)	10^6–10^{16}	10^4–10^{11}	10^4–10^8
Cell distribution	Uniform distribution	Generally uniform but sometimes the cells in the core are different in size from those found at the edges	It is difficult to obtain foams with uniform cell
Surface quality	Good	Good and glossy	Generally poor
Skin layer thickness (μm)	Thin	Thin	Thick
Addition of nucleating agents/process flexibility	Foaming composition is fixed from the onset. Must be done in the previous processes such as injection molding or extrusion, etc.	Composition can be changed at any time. Nucleating agents can be introduced at anytime during processing	Nucleating agents can be introduced also at anytime during processing
Blowing agent supply	Sample is saturated with the blowing agent until equilibrium is reached	Blowing agent is metered but not more than the melt can take	Blowing agent is metered but not more than the melt can absorb at a certain processing condition
Tooling cost	Cheaper than the others	Expensive depending on the machine capacity	Expensive depending on machine capacity and also mold is extra cost

Table 1. Comparison on batch foaming, foam extrusion, and foam injection molding.

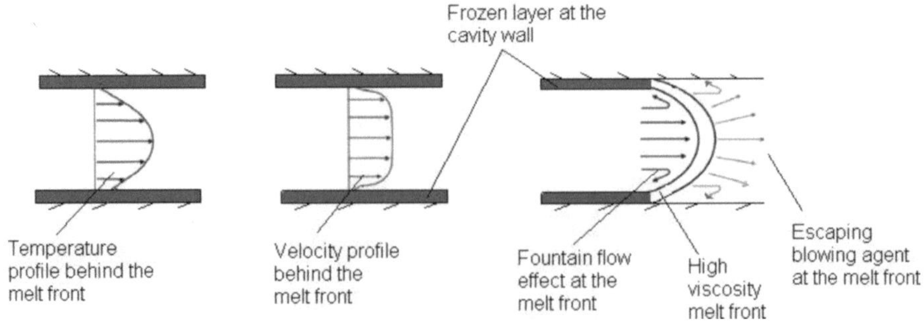

Figure 7. Illustration of the headstream in foam injection molding.

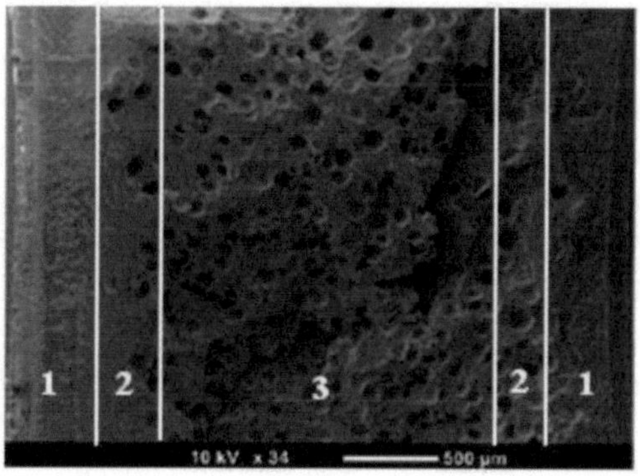

Figure 8. Zones of the injected molded foams according to the cell morphology (1) compact skin layer, (2) outer foam core, (3) inner foam core [20].

has critical effect in improving the properties of the polymer foam. Besides all, generation of the foam cells is effective in reducing the sink marks, warpage internal stresses, and shrinkage of the foam plastic, which enhance the selection of the foam injection process in the plastic industry.

3. Thermoplastic foams: processing and nanocomposites

A wide variety of thermoplastics such as polypropylene, polyethylene, polystyrene, polycarbonate, polyvinyl chloride, polylactic acid, and polycarbonate have been experienced in foam processing techniques. Depending on their viscosity, melt strength, the formation of cell morphology of the polymer foam changes. Due to the demands of improvements in foam morphology and the mechanical strength of the polymer foams, nanoparticle-reinforced polymer nanocomposites have been developed in last decade. It has known that nanoparticle usage in polymer foam processing improved cell morphology due to the nucleating agent behavior of the nanoparticles in the polymer matrix. The presence of nanoparticles is also effective in improving the mechanical, physical, and chemical properties of polymer foams. In this section, the most experienced thermoplastic foams in industrial applications and their composites are reviewed.

3.1. Polypropylene-based foams

Polypropylene, member of linear polyolefin group, has poor solubility of carbon dioxide and low melt strength. Linear olefins do not show high strain–induced hardening which is the critical requirement in withstanding the stretching force generated in the stages of cell growth. Chien et al. [21] studied on polypropylene foams obtained by conventional injection molded and traditional foaming injection molded using chemical foaming agent content under various molding conditions. They observed the effects of process parameters, part thickness and foaming agent content on the degree foaming. Injection velocity, melt temperature, mold temperature, and back pressure on weight reduction and mechanical properties were investigated. The chemical foaming agent was azodicarbonamide used in their study. It has been reported higher injection velocity induced higher weight reduction due to the reduction in the amount of melt foaming in the screw and provided more melt foaming in the cavity. Higher melt temperature and mold temperature resulted in higher melt foaming in the cavity; consequently, weight reduction was observed. The effect of foaming agent content on weight reduction of thick parts was evaluated, and it was found that concerned the weight decreases with increasing content of foaming agent but less significantly. The mechanical test results of PP foam showed that tensile, flexural strength, stiffness, and part weight decreased with the increasing melt temperature, mold temperature, and injection velocity whereas increased with increasing back-pressure.

Sporrer and Altstadt [19] obtained PP foams by physical foaming, MuCell technique. The effect of process conditions on cell morphology was observed. Two different mold temperatures were studied as 20 and 80°C. When they worked at higher mold temperatures, the thickness of the compact skin layers was reduced by 20% as compared to the part processed using the cold mold. The SEM image is given in **Figure 9**. The mold with 80°C gave 552 μm of layer thickness, and the mold with 20°C gave 442 μm of skin layer. The thinner skin layer is the result of the lower thermal gradient between melt and mold steel and a less rapid heat transfer in hotter mold.

In **Figure 9**, morphologies of polypropylene foams are given that are foam injected molded under 20 and 40°C. The foam of 40°C gave coarser and ruptured cells, while the foam of 20°C gave thicker skin layer. The reason of thicker skin layer is the sudden frozen layer of the material as it is injected into the cold mold wall (20°C).

Xin et al. [22] applied chemical foaming by using azodicarbonamide in order to obtain microcellular polypropylene/waste rubber tire (WGRT). Their aim was to generate "a value added" product by using a waste material. They observed the effects of critical processing parameters on cell morphology and physical properties of the blend foams. They observed that under the

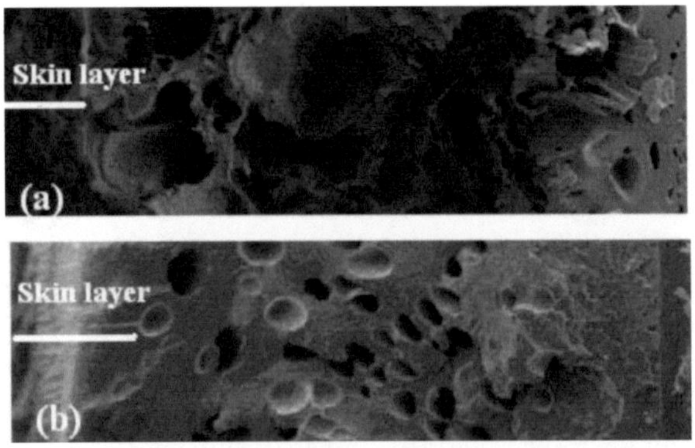

Figure 9. Morphology of the PP foams processed with different mold temperatures (a) 40°C (b) 20°C [20].

same molding conditions, the microcellular PP/WGRT blend samples had smaller cell sizes and higher cell densities than the microcellular PP samples. They reported that it was due to the behavior of the waste rubber tire powders as nucleating agent that promoted heterogeneous cell nucleation, resulting in higher cell density. On the other hand, the increase of viscosity in the PP/WGRT blend prevented the growth of the cells, leading to a smaller cell size [23].

Realinho et al. [24], developed flame-retardant polypropylene composite foams by combining a basic hydrated magnesium carbonate (hydromagnesite), an intumescent additive based on ammonium polyphosphate, an organo modified-montmorillonite (MMT), and graphene nanoplatelets with PP. Azodicarbonamide was used in chemical foaming. Addition of hydromagnesite was 60%, while the other nanoparticles were about 1%. They reported that cell size reduced to 100 μm from 900 μm with the addition of hydromagnesite. The presence of nanoparticles enhanced the cell morphology. They also mentioned that solid composites were more successful in improving the flame retardancy than foam composites.

In order to improve mechanical properties of PP foams, Hwang and Hsu [25] used nanosilica particle polypropylene. Physical foaming, MuCell technique, was applied in their study. Particle addition was between 2 and 10%. They observed that when the silica content increased, cell size decreased and the cell density increased. However, a threshold was seen in silica content that the cell size leveled off when the nanosilica loading was greater than 4%. Similarly to the previous studies, dispersion of the nanoparticles in the matrix homogenously improved cell morphology. This is due to the nucleating agent effect of nanoparticles that cells nucleate at the boundary between the polymer matrix and the filler. Hwang and Hsu [25] also experienced the effect of microsilica particles and compare their effect on cell generation. They observed that at the same concentrations of particles, nanoparticles gave denser and smaller cells in size.

Nanoclay is another nanoparticle used in order to improve the properties of polypropylene foams. Nanoclay particles, similar to silicate, act as nucleating agent and lead homogeneity in cell size. Increment in clay content decreased the cell size due to the high viscosity of polymer [16, 26, 27]. Furthermore, authors suggested that clay particles act as secondary layer to protect the cells from being destroyed by external forces. In other words, the biaxial flow of the material during foam processing, the nanoparticles align along the flow direction which is the cell boundary (**Figure 10**). By this way, clay particles help the cells withstand the stretching force. Otherwise, the cell wall will break and weaken the mechanical strength of the polymer foam.

Doruk [28] studied on the effects of the nanocalcite and microcalcite particles on the cell morphology and mechanical strength of the PP foams. Nanoparticles were mixed with polymer in twin screw extruder, and then, foam injection molding was applied by chemical foaming agent (azodicarbonamide). When the fracture surface was observed as given in **Figure 11**, nanoparticle addition improved cell morphology. In **Figure 12**, tensile properties of the PP/calcite foams are given, and it has been seen that under the same concentration of the particle addition (1 wt.%), tensile strength of PP/microcalcite foam is slightly higher than that of PP/nanocalcite foam. This is due to the improved cell generation of the PP/nanocalcite foam as given in **Figure 11**. On the other hand, cell generation of PP/microcalcite is very poor, and the ductility of the PP/nanocalcite is apparently higher than that of PP/micro. When the weight

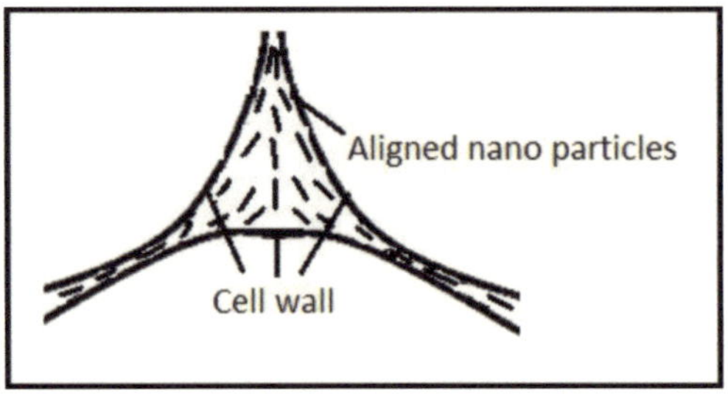

Figure 10. Illustration of the alignment of the nanoparticles during foaming process.

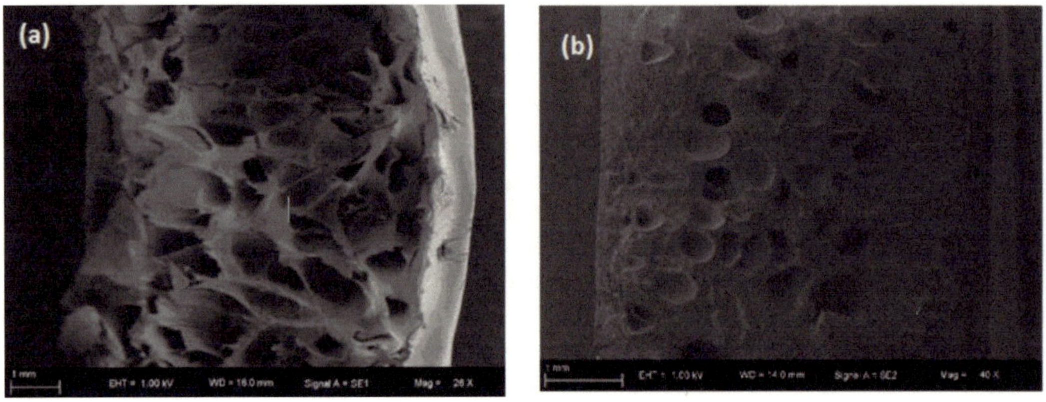

Figure 11. Cell morphology of PP/calcite foams (a) nanocalcite reinforced (b) microcalcite reinforced [28].

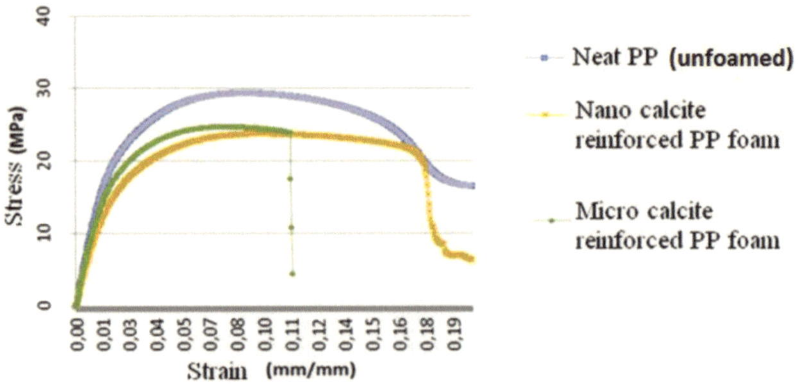

Figure 12. Comparison tensile properties of PP foams with microsized and nanosized calcite (1 wt.%) [28].

loss is considered, nanocomposite foam shows the weight loss of 20.7%, while microcomposite foams have 8.3% of weight loss.

The demand of new lightweight materials with improved transport properties for applications in electrostatic discharge, fuel system components, and electromagnetic interference shielding such as fuel cells, gaskets for electronic devices, among others brings the generation of a multi-functional material, carbon-based nanoparticle-reinforced polymer foams. Carbon nanotubes, graphene, have been recently attractive for many applications in electronic industry. Antunes et al. used carbon nanofibers (CFN) with polypropylene in order to improve thermal and electrical properties of the polypropylene composite foams [29, 30]. In their study, they emphasized the importance of particle alignment during cell generation and the importance of this on thermal conductivity of the PP. The foaming of PP with CNF provided a kind of network of the particles through the polymer matrix which increased the thermal conductivity of the polymer. When they compared their results with foamed and unfoamed polymer composites, they observed that the unfoamed composite showed a constant thermal conductivity independently of CFN content, while PP/CFN foams showed an increment in thermal conductivity as the content of CFN increased. This shows that a kind of network of CNFs throughout the polymer was formed formation that makes the material thermally conductive. The formation of this network is similar to the clay alignment as discussed in **Figure 10**. In a different study related to PP/CNF foams [31], electrical conductivity of the polymer composite foams was investigated. When the unfoamed and foamed composite were compared, the lower concentration of CFN gave high electrical conductivity. Also, the cellular structure generated during processing with cells highly elongated along the foam thickness direction increased the through-plane electrical conductivity of the foams with regard to the in-plane one. This indicates the importance of cell morphology on the electrical properties of the polymer foams. An accurately developed cellular structure may help to develop foams for semi-conducting lightweight materials [29–31].

Altan [20] made a research on polypropylene/ nano–zinc oxide (ZnO) foams. Zinc oxide is another alternative material to improve electrical properties of the polymer foams. The concentration of ZnO was 1.5% in weight. When the cell morphologies of PP foam and PP/nano-ZnO foam were compared, it has been seen that the presence of nanoparticles decreased the cell diameter and thickness of skin layer and increased the cell dense (**Figure 13**).

Graphene is the latest nanomaterial applied in polymer foams. Similar to previous nanofillers, in literature, it has been seen that graphene loading to PP between 2.5 and 5 wt.% has great

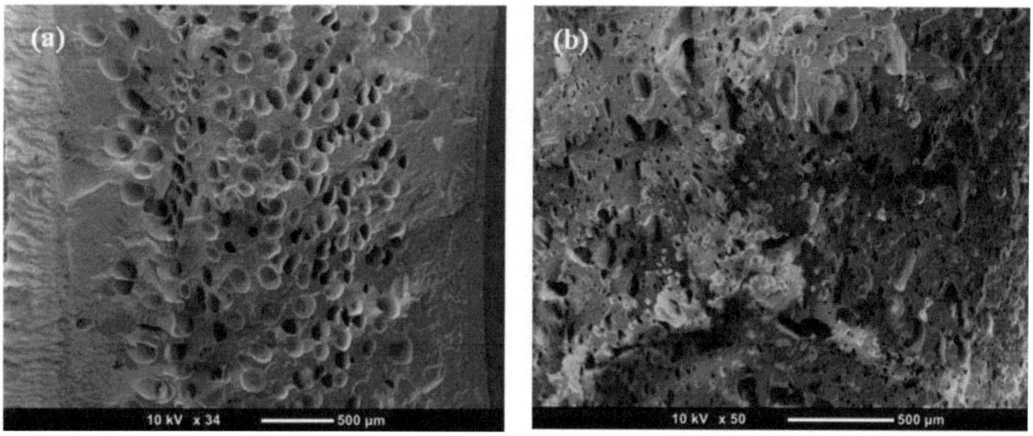

Figure 13. Fracture surfaces of polypropylene foams (a) neat PP (b) PP/ZnO [20].

effect in cell nucleation [32]. Besides, the higher expansion of polymer during foaming process induces higher exfoliation of the graphene nanoplatelets in PP matrix and brings higher mechanical strength [32].

3.2. Polyethylene-based foams

Polyethylene (PE) is a member of polyolefin-like polypropylene. High-density polyethylene (HDPE) and low-density polyethylene (LDPE) have been experienced in foam processing. LDPE foams are used as thermoplastic material for applications such as packaging and foamed sheets, sports parts due to its owing low density, high elasticity, water resistance, and low cost. One of the common problems in polymer foams is the loss of toughness and ductility of the material due to the generation of the cells. Sun et al. [33] developed a toughening mechanism for high-density polyethylene/polypropylene blends. They obtained super ductile polymeric blends using microcellular injection molding. They prepared PP/HDPE and PP/LDPE blends which were prepared at weight ratios of 75/25, 50/50, and 25/75 with melt mixing method and then by applying MuCell technique. It has been observed that during tensile test, 75/25 PP/LDPE foamed parts were highly fibrillated along the tensile load direction in the necking region. The researchers reported that the reason of this behavior—the high ductility of polymer foams—was related to two reasons. The first one was due to the microcellular foam structure cell size lower than 100 μm, and the other was an immiscible but compatible submicron-size secondary polymeric phase. During tensile test, the sub-micron phase of the blend debonds from the matrix, and the cavities collapse. Secondly, they interconnect the microscale foam cells along the load direction. This generates many fibrils that make the material highly ductile [33].

Similar to the case of PP-nanocomposites foams, various authors have reported the preparation, characterization, and properties of PE-nanocomposite foams [34–37]. Arroyo et al. [37] developed low-density polyethylene/silica nanocomposite foams by using chemical foaming agent. They applied different concentrations of silica between 1 and 9%, and foaming agent was 5% in weight. The addition of silica particles improved the cellular structure of the LDPE improved with increment of the cell densities and decrease in cell size. However, at silica concentrations over 6%, increase in cell size was reported. There are few reasons about the poor quality of foam cell morphology under higher nanoparticle concentrations. One of them is possible agglomerations of the nanoparticles at higher concentrations that they prevent the formation of the cells. Also, increase in viscosity of the polymer melt because of the higher loadings of the particles makes the cell generation difficult.

Clay is one of the most used inorganic particles in enhancing the properties of PE-based foams. Clay, such as montmorillonite (MMT), is mixed with polymers and the mechanical strength of the polymers increase [36, 38]. In the study of Hwang et al. [38], the effect of MMT on the cell morphology of the low density of polyethylene (LDPE) was observed. First of all, the researchers enhanced the nanoparticle distribution in the polymer matrix by grafting polar maleic anhydride (MA) onto nonpolar LDPE. The concentration of MMT was between 1 and 5%. Their results are similar to the previous studies that the MMT and MA act as nucleating agents that lead to a finer and more uniform cell structure. When the dispersion of the nanoparticles is homogenous, the cell size decreases and the distribution of the cells is homogeneous.

Polyethylene foams, similar to the other thermoplastic foams, can be processed either batch foaming or foam injection molding. Hayashi et al. [39] compared the orgona clay PE-based ionomer composite foams obtained by batch processing and foam injection foaming. The effect of clay on foam morphology of PE is similar to the previous studies that the dispersed nanoclay particles act as nucleating sites for cell formation, and cell growth occurs on the surfaces of the clays. Different from batch processing, in foam injection molding, the moldings have two compact solid skin layers and a foamed core. In both foaming processes, the foam morphology can be improved by setting process conditions correctly depending on the viscosity of the polymer and the temperature and gas pressure limits. Hayashi et al. [39] reported that in batch process, the ionic cross linked structure provided finer cells, and the coalescence of the cells was prevented. On the other hand, by the effect of supercritical nitrogen gas as foaming agent during foam injection molding process, the viscosity of polymer was decreased, and this promoted the nucleation and also coalescence of the cells, especially at high temperatures.

3.3. Polystyrene-based foams

Polystyrene (PS) is an amorphous polymer, and it has wide application area in polymer foam processing such as thermal insulation, packing material due to its low cost, ease of processing, resistance to moisture, and recyclability. Dow Chemical Company invented PS foams as "Styrofoam" in 1941. Polystyrene foams are basically divided into two; expanded polystyrene (EPS) and extruded polystyrene (XPS). Expanded polystyrene has white color and can be used in cups for hot beverages, insulation material in whitegoods, or in packing industry. EPS consists of 96–98% air and 2–4% polystyrene. Processing method is heating the material with steam and then expanding of the material. Extruded polystyrene (XPS) has smaller air pockets inside and manufactured by extrusion process in the form of boards with different colors to identify the brand type of the product. Zhang et al. [40] produced extruded polystyrene foams (XPS) by using CO_2 and water as a co-blowing agent. Okolieocha et al. [41] carried out on a tandem foam extrusion line, similar studies on XPS. They used a slit die (0.5 mm) set at a temperature of 126°C. In order to enhance cell density, they used 1 wt.% of thermally reduced graphite oxide. However, general purpose polystyrene (GPSS) and high-impact polystyrene (HIPS) are suitable for injection molding and structural foaming, and cell generation can be provided similar to the other thermoplastics by chemical or physical foaming agents. Furthermore, PS is not characterized by low melt strength so this makes it suitable for foam injection molding. Hwang et al. [42] applied foam injection molding via MuCell for obtaining clay-reinforced PS foams. Clay was used to improve cell morphology of the polystyrene foams. They obtained PS/clay composite foams with small size cell, which makes the material highly suitable for acoustic and thermal insulation applications. On the other hand, layers of clay-like montmorillonite (MMT) are difficult to fully exfoliate in PS matrix. MMT was modified by stearylbenzyldime-ammonium chloride prior melt mixing with polystyrene, and the concentration of MMT in the matrix was experienced in a narrow range as 0.25-0.5-1-2-3% (wt). It has been seen that organo clay presence of 1% in PS matrix gave the small cell in diameters, leading to maximum tensile strength, thermal stability, and cell density.

3.4. Polylactic acid–based foams

Poly (lactide acid) or polylactide (PLA) is a biodegradable and biocompatible polymer produced from such renewable sources as cornstarch and sugarcane [1–4]. PLA foam is a competitive material among most of other thermoplastic foams due to its biocompatibility and biodegradability, PLA has been widely used in tissue engineering applications such as skin, bones, blood vessels, due to their highly porous structure as scaffolds in last [4]. The porous surface of the PLA foams enhances the biological activities of both seeded and native cells. High porosity is important for enhancing biological properties of the scaffold such as the adhesion, proliferation, and migration of the cells. However, mechanical properties of foams decrease with the increasing of porosity. Besides, the high strength and brittle properties of PLA make it difficult to use and process it in foaming techniques. Researchers are focusing on generation of PLA with different polymers or PLA matrix composites [4].

Similar to the other thermoplastics, PLA foams with uniform cell morphology are generally obtained by physical foaming agents such as carbon dioxide and nitrogen in foam injection molding and foam extrusion. However, the poor melt strength of PLA brings difficulties in obtaining an enhanced cell morphology. There are several ways to improve PLA's foam morphology by means of improving melt strength of the polymer such as using chain extenders, using polymer blends of PLA, addition of nanoparticles, and improving crystallization kinetics. Low melt strength of PLA induces cell coalescence during cell growth. Addition of chain extenders to PLA increased the rheological properties of PLA, and depending on this, cell morphology is enhanced [43–45].

Crystallization is an important factor in improving melt strength and foaming ability of thermoplastics. The low melt strength of PLA can be promoted by improving crystallization kinetics and the poor viscoelastic behavior of the polymer. However, high crystallinity has negative effect on cell generation by suppressing the foam expansion. On the other hand, during foaming, cell nucleation starts around crystals [46, 47]. Therefore, improving crystallinity can be balanced by some nucleating agents such as additives and nanofillers behave-like nucleating agents. There are several studies on PLA nanocomposite foams that used calcite, sepiolite, and multi-walled carbon nanotube as nanofiller [46–49]. In these studies, nanomaterial addition was found to be nucleating agent for crystallinity and cell generation. There has been great interest to clay-reinforced PLA composite foams due to the enhanced viscoelastic behavior of clay particles in the polymer matrix which improves cell morphology [48, 50]. As the nano clay particles increased, the cell density of the foamed samples increased. It has been reported that even a small amount addition of carbon nanotube (CNT) promoted the cell density due to its effect on cell nucleation [47]. An interesting point about PLA/CNT composite foams that the gas used during foam injection molding behaved like a dispersant for the nanoparticles that homogenous dispersion of CNT could be obtained in the polymer matrix. This is due to the plasticization effect of the supercritical fluid phase of CO_2 [43, 47]. Therefore, in foam extrusion and foam injection molding, the foaming agents do not only provide foaming but also disperse the particles homogenously in the matrix.

http://dx.doi.org/10.5772/intechopen.71083

4. Conclusion

Thermoplastic foams are generally obtained by batch foaming, foam extrusion, and foam injection foaming. Batch foaming is cheaper than the others due to simple equipment, but in each method, the main aim is to promote cell morphology by providing small cell in diameter and high cell density in the polymer matrix. The thermal properties of the polymer, its viscosity, degree of crystallinity, and melt strength are the important factors in improving the cell morphology. There are several ways to improve the cell morphology of the thermoplastics such as preparation of polymer blends, using chain extenders or using nanofillers. Nanofiller addition is popular in last decade due to improvements in properties of the polymer foams. It is known that some nanoparticles are difficult to disperse in the polymer matrix because they tend to agglomerate seriously. However, in polymer foam processing, usage of foaming agent such as CO_2 or N_2 gases enhances the dispersion of the particles by reducing providing plasticization effect. The homogenous distribution of the nanoparticles contributes the cell nucleation.

Nanocalcite, nanomontmorillonite, nanosilicate, and carbon nanotube are the most used nanoparticles in polymer foams. Graphene-reinforced polymer foams are still under investigation. Both carbon nanotube and graphene-reinforced polymer foams have application area as thermal insulator or electrical conductive polymer foams. Nanocalcite or nanosilicate has been used for improving cell generation, increasing mechanical strength, and enhancing flame retardancy of the polymer foam. It has been seen that small amount of nanofiller addition improved cell morphology seriously.

Polypropylene and polystyrene foams are rigid foams that have wide application area in automotive and wind industries. On the other hand, polylactic acid is a promising biomaterial, and PLA foams are suitable materials for tissue engineering as scaffolds. The high porosity of the PLA foams, as scaffolds, provides enhanced biological activities of both seeded and native cells, and they can substitute the native tissue until the native tissue heals.

Acknowledgements

The author would like to thank The Scientific and Technological Research Council of Turkey (TUBITAK) for its support on the research about polypropylene/zinc oxide nanocomposite foams, which was carried under International Post Doctoral Research Fellowship Program-2219.

Author details

Mihrigul Altan

Address all correspondence to: meksi@yildiz.edu.tr

Department of Mechanical Engineering, Yildiz Technical University, Istanbul, Turkey

References

[1] Ibeh CC, Lee SW. Current trends in nanocomposite foams. Journal of Cellular Plastics. 2008;**44**:493-515. DOI: 10.1177/0021955X08097707

[2] Lee LJ, Zen C, Cao X, Han X, ShenJ XG. Polymer nanocomposite foams. Composites Science and Technology. 2005;**65**:2344-2363. DOI: 10.1016/j.compscitech.2005.06.016

[3] Zhai W, Park CB, Kontopoulou M. Nanosilica addition dramatically improvesthe cell morphology and expansion ratio of polypropylene heterophasic copolymer foams blown in continuous extrusion. Industrial and Engineering Chemical Research. 2011;**50**: 7282-7289. DOI: 10.1021/ie102438p

[4] Mi HY, Salick MR, Jing X, Jacques BR, Crone WC, Peng XF, Turng LS. Characterization of thermoplastic polyurethane/polylactic acid (TPU/PLA) tissue engineering scaffolds fabricated by microcellular injection molding. Materials Science and Engineering. 2013;**33**:4767-4776. DOI: 10.1016/j.msec.2013.07.037

[5] Chaudhary AK, Jayaraman K. Extrusion of linear polypropylene clay nanocomposite foams. Polymer Engineering and Science. 2011;**51**:1749-1756. DOI: 10.1002/pen.21961

[6] Roberts RD, Kwok JC. Styrene-maleic anhydride copolymer foam for heat resistant packaging. Journal of Cellular Plastics. 2007;**43**:135-143. DOI: 10.1177/0021955X06074329

[7] Sorrentino L, Aurilia M, Iannace S. Polymeric foams from high performance thermoplastics. Advances in Polymer Technology. 2011;**30**:234-243. DOI: 10.1002/adv.20219

[8] Subramonian S, Filiccia P. Alcott: Novel soft touch, low abrasion, fine cell polyolefin foams. Journal of Cellular Plastics. 2007;**43**:331-343. DOI: 10.1177/0021955X07079070

[9] Liu S, Duvigneau J, Vancso GS. Nanocellular polymer foams as promising high performance thermal insulation materials. European Polymer Journal. 2015;**65**:33-45. DOI: 10.1016/j.eurpolymj.2015.01.039

[10] Okolieocha C, Raps D, Subramaniam K, Altstad V. Microcellular to nanocellular polymer foams: Progress (2004-2015) and future directions – A review. European Polymer Journal. 2015;**73**:500-519. DOI: 10.1016/j.eurpolymj.2015.11.001

[11] Nalawade SP, Picchioni F, Janssen LPBM. Supercritical carbon dioxide as a green solvent for processing polymer melts: processing aspects andapplications. Progress in Polymer Science. 2006;**31**:19-43. DOI: 10.1016/j.progpolymsci.2005.08.002

[12] Sauceau M, Fages J, Common A, Nikitine C, Rodier E. New challenges in polymer foaming: A review of extrusion processes assisted by supercritical carbon dioxide. Progress in Polymer Science. 2011;**36**:749-766. DOI: 10.1016/j.progpolymsci.2010.12.004

[13] Ruiz JAR, Vincent M, Agassant JF, Sadik T, Pillon C. Polymer foaming with chemical blowing agents: Experiment and modeling. Polymer Engineering and Science. 2015;**55**:2018-2029. DOI: 10.1002/pen.24044

[14] Rohleder M, Bledzki AK, Kirschling H. Correlation between injection moulding parameters, morphology and properties of microcellular polycarbonate. Journal of Cellular Plastics. 2012;**48**:291-300. DOI: 10.1177/0021955X12449218

[15] Zhao G, Wang G, Wang J, Zhang L. Foaming mechanism and control in microcellular injection molding. Society of Plastic Engineers Plastic Research Online. 2014:1-4. DOI: 10.2417/spepro.005571

[16] Rohleder M, Jakob F. Specialized Injection Molding Techniques. USA: Elsevier; 2016. 262 p. DOI: 10.1016/B978-0-323-34100-4.00002-X

[17] Villamizar CA, Han CF. Studies on structural foam processing, the rheology of foam extrusion. Polymer Engineering and Science. 1978;**18**:687-698. DOI: 10.1002/pen.760180904

[18] Gosselin R, Rodrigue D. Cell morphology analysis of high density polymer foams. Polymer Testing. 2005;**24**:1027-1035. DOI: 10.1016/j.polymertesting.2005.07.005

[19] Sporrer A, Altstadt V. Controlling morphology of injection molded structural foams by mold design and processing parameters. Journal of Cellular Plastics. 2007;**43**:313-330. DOI: 10.1177/0021955X07079043

[20] Altan M. Development of polymer composite foams by injection molding using chemical foaming agents. International Post Doctoral Research Fellowship Programme (Final Report), TUBITAK. 2016

[21] Chien RD, Chen SC, Lee PH, Huang JS. Study on the molding characteristics and mechanical properties of injection-molded foaming polypropylene parts. Journal of Reinforced Plastics and Composites. 2004;**23**:429-444. DOI: 10.1177/0731684404031891

[22] Xin ZX, Zhang ZX, Pal K, Byeon JU, Lee SH, Kim JK. Study of microcellular injection-molded polypropylene/waste ground rubber tire powder blend. Materials & Design. 2010;**31**:589-593. DOI: 10.1016/j.matdes.2009.07.002

[23] Zhang ZX, Lee SH, Kim JK, Zhang SL, Xin ZX. Preparation and characterization of polypropylene/waste ground rubber tire powder microcellular composites by supercritical carbon dioxide. Macromolecular Research. 2008;**16**:404-410

[24] Realinho V, Haurie L, Antunes M, Velasco JI. Thermal stability and fire behaviour of flame retardant highdensity rigid foams based on hydromagnesite-filled polypropylene composites. Composites: Part B. 2014;**58**:553-558. DOI: 10.1016/j.compositesb.2013.11.015

[25] Hwang S, Hsu PP: Effects of silica particle size on the structure and properties of polypropylene/silica composites foam. Journal of Industrial Engineering and Chemical Research. 2013;**19**:1377-1383. DOI: https://doi.org/10.1016/j.jiec.2012.12.043

[26] Nam PH, Maiti P, Okamoto M, Kotaka T, Nakayama T, Takada M. Foam processing and cellular structure of polypropylene/clay nanocomposites. Polymer Engineering and Science. 2002;**42**:1907-1918. DOI: 10.1002/pen.11083

[27] Okamoto M, Nam PH, Maiti P, Kotaka T, Nakayama T, Takada M, et al. Biaxial flow-induced alignment of silicate layers in polypropylene/clay nanocomposite foam. Nano Letters. 2001;**1**:503-505. DOI: 10.1021/nl010051+

[28] Doruk G. Improvement properties of polymer foams obtained by foam injection molding [thesis]. Istanbul: Yildiz Technical University; 2017

[29] Antunes M, Realinho V, Solórzano E, Rodríguez-Perez MA, Saja JA, Velasco JI. Thermal conductivity of carbon nano fibre polypropylene composite foams. Deffect and Diffusion Forum. 2010;**297-301**:996-1001. DOI: 10.4028/www.scientific.net/DDF.297-301.996

[30] Antunes M, Mudarra M, Velasco JI. Broad-band electrical conductivity of carbon nanofibre-reinforced polypropylene foams. Carbon. 2011;**49**:708-717. DOI: 10.1016/j.carbon.2010.10.032

[31] Antunes M, Velasco JI, Realinho V, Arencón D. Characterization of carbon nanofibre-reinforced polypropylene foams. Journal of Nanoscience Nano Technology. 2010;**10**: 1241-1250. DOI: 10.1166/jnn.2010.1831

[32] Verdejo R, Barroso-Bujans F, Rodríguez-Perez MA, Saja JA, Lopez-Manchado MA. Functionalized graphene sheet filled silicone foamnanocomposites. Journal of Materials Chemistry. 2008;**18**:2221-2226. DOI: 10.1039/B718289A

[33] Sun X, Kharbas H, Peng J, Turng LS. Fabrication of super ductile polymeric blendsusing microcellular injection molding. Manufacturing Letters. 2014;**2**:64-68 https://doi.org/10.1016/j.mfglet.2014.02.002

[34] Velasco JI, Antunes M, Ayyad O, Lopez-Cuesta JM, Gaudon P, Saiz-Arroyo C. Foaming behaviour and cellular structure of LDPE/hectorite nanocomposites. Polymer. 2007; **48**:2098-2108. DOI: 10.1016/j.polymer.2007.02.008

[35] Gutierrez EL, Arroya CS, Velasco JI, Perez MAR: Low density polyethylene/silica nanocomposite foams. Relationship between chemical composition, particle dispersion, cellular structure and physical properties. European Polymer Journal. 2016;**81**:173-185. DOI: 10.1016/j.eurpolymj.2016.06.001

[36] Jin DW, Seol SM, Kim GH. New compatibilizer for linear low-density polyethylene (LLDPE)/clay nanocomposites. Journal of Applied Polymer Science. 2009;**114**:25-31. DOI: 10.1002/app.30544

[37] Arroyo CS, Perez MAR, Velasco SJA. Influence of foaming process on the structure-properties relationship of foamed LDPE/silica nanocomposites. Composites Part B Engineering. 2013;**48**:40-50. DOI: 10.1016/j.compositesb.2012.10.045

[38] Hwang S, Peming P, Yeh HJ, Yang J, Chang K, Lai YZ. Effect of clay and compatibilizer on the mechanical/thermal properties of microcellular injection molded low density polyethylene nanocomposites. International Communications in Heat and Mass Transfer. 2009;**36**:471-479. DOI: 10.1016/j.icheatmasstransfer.2009.02.005

[39] Hayashi H, Mori T, Okamoto M, Yamasaki S, Hayami H. Polyethylene ionomer-based nano-composite foams prepared by a batch process and MuCell injection molding. Materials Science and Engineering C. 2010;**30**:62-70. DOI: 10.1016/j.msec.2009.08.009

[40] Zhang C, Zhu B, Lee LJ. Extrusion foaming of polystyrene/carbon particles using carbon dioxide and water as co-blowing agents. Polymer. 2011;**52**:1847-1855

[41] Okolieocha C, Koppl T, Kerling S, Tolle FJ, Fathi A, Mülhaupt R, Altstadt V. Influence of graphene on the cell morphology and mechanical propertiesof extruded polystyrene foam. Journal of Cellular Plastics. 2015;**51**:413-426. DOI: 10.1177/0021955X14566084

[42] Hwang S, Peming PH, Yeh J, Hu C, Chang KC. Effect of organoclay on the mechanical/thermal properties of microcellular injection molded polystyrene-clay nanocomposites. International Communications in Heat and Mass Transfer. 2009;**36**:799-805. DOI: 10.1016/j.icheatmasstransfer.2009.06.011

[43] Nofar M, Park CB. Poly (lactic acid) foaming. Progress in Polymer Science. 2014;**39**:1721-1741. DOI: 10.1016/j.progpolymsci.2014.04

[44] Ameli A, Jahani D, Nofar M, Park CB. Development of high void fraction polylactide composite foams using injection molding: Mechanical and thermal insulation properties. Composites Science and Technology. 2013;**90**:88-95. DOI: doi.org/10.1016/j.compscitech.2013.10.019

[45] Corre YM, Maazouz A, Duchet J, Reignier J. Batch foaming of chain extended PLA with supercritical CO_2: Influence of the rheological properties and the process parameters on the cellular structure. Journal of Supercritical Fluids. 2011;**58**:177-188. DOI: 10.1016/j.supflu.2011.03.006

[46] Liao X, Nawaby AV, Whitfield PS. Carbon dioxide-induced crystallization in poly(l-lactic acid) and its effect on foam morphologies. Polymer International. 2010;**59**:1709-1718. DOI: 10.1002/pi.2910

[47] Antunes M, Velasco JI. Multifunctional polymer foams with carbon nanoparticles. Progress in Polymer Science. 2014;**39**:486-509. DOI: 10.1016/j.progpolymsci.2013.11.002

[48] Di Y, Iannace S, Di Maio E, Nicolais L. Poly(lactic acid)/organo clay nanocomposites: thermal, rheological properties and foam processing. Journal Polymer Science Part B: Polymer Physics. 2005;**43**:689-698. DOI: 10.1002/polb.20366

[49] Najafi N, Heuzay MC, Carreau PJ, Therriault D, Park CB. Mechanical and morphological properties of injection molded linear and branched-polylactide (PLA) nanocomposite foams. European Polymer Journal. 2015;**73**:455-465. DOI: 10.1016/j.eurpolymj.2015.11.003

[50] Fujimoto Y, Ray SS, Okamoto M, Ogami A, Yamada K, Ueda K. Well-controlled biodegradable nanocomposite foams: From microcellular to nanocellular. Macromolecular Rapid Communication. 2003;**24**:457-461. DOI: 10.1002/marc.200390068

Chapter 7

PRAP-CVD: A Novel Technique to Deposit Intrinsically Conductive Polymers

Bianca Rita Pistillo, Kevin Menguelti, Didier Arl,
Renaud Leturcq and Damien Lenoble

Additional information is available at the end of the chapter

http://dx.doi.org/10.5772/intechopen.71736

Abstract

Polymers provide extraordinary opportunities for functionalizing surfaces integrated into flexible devices contributing to a significant advancement in thin-film technologies. Both the advantageous characteristics of conventional polymers (e.g. low weight, flexibility) and the functional physical properties of conventional semiconductors (e.g. absorption and emission of light and a tuneable conductivity) can be found in polymers providing innovative materials. Among polymers with heterocyclic structures, called conjugated polymers, polythiophene remains one of the most intensely researched materials in the field of so called organic electronics owing to the relatively facile and well-established synthetic modifications of the corresponding monomers and its derivatives. In particular, poly(3,4-ethylenedioxythiophene) (PEDOT) is one of the most promising owing to its exceptional stability, transparency, and electrical conductivity. Nevertheless it is difficult to process PEDOT into thin-films by traditional solution-based methods. Plasma Radicals Assisted Polymerisation via Chemical Vapour Deposition (PRAP-CVD) is a novel technique able to overcome the challenges caused by the conventional techniques.

Keywords: chemical vapour deposition, conjugate polymers, poly(3,4-ethylenedioxythiophene), conductivity, organic electronics

1. Introduction

Considering the early days of conjugated and conducting polymer synthesis, the first works on polypyrrole as well as polyaniline are often referred to as landmark developments in this field [1, 2]. Despite these polymers resulted completely insoluble and infusible as formed from oxidative polymerisation, they served as basis for inducing electroactivity into polymer systems. The Nobel Prize winners, Heeger, MacDiarmid and Shirakawa, have the merit of discovering that the treatment of these polymers with controlled amounts of halogens could induce conductive behaviour, demonstrating for the first time conductive properties of an organic polymer [3–5]. They were the pioneers of the investigation of the mechanisms and physical origins underlying the charge transport in this new class of materials, starting by studying polyacetylene [6].

A few years later, in 1980, the discovery of electrical conductivity in oxidised polythiophenes was reported respectively by Yamamoto and Lin [7, 8]. The mechanism responsible of polythiophene conductivity was found to be similar to that one of oxidatively doped polyacetylene, i.e. generation of delocalized radicals by an oxidising dopant, and subsequent stabilisation via ionic interactions between the charged polymer and the spent dopant. Although the first observations recorded low conductivity values (0.001–0.1 S/cm) compared to doped acetylene, thiophene monomers could offer a more diverse opportunity for functionalization than bare acetylene owing to their chemical structure. Since the evolution of conjugated polymers and their use as semiconductors, polythiophene and its derivatives have remained one of the most intensely researched materials in the field of so called organic electronics. Polythiophenes have been vastly explored in a variety of applications such as organic field-effect transistors (OFETs), organic light-emitting diode (OLEDs), organic photovoltaics (OPVs), and sensing devices in medical and biological fields [9–11].

The studies of polythiophene derivatives culminated with the synthesis of Poly(3,4-ethylenedioxythiophene) (PEDOT) by Bayer in 1988 [12]. PEDOT or PEDT, belongs to the moderate amount of conductive polymers, which have not only attracted remarkable scientific interest but also companies since it is currently used as material in different products of modern life. When PEDOT was invented in 1988, a so huge number of its potential applications was not really realised at that time. Since then this has dramatically changed, and nowadays PEDOT is probably the best conducting polymer available in terms of conductivity, processability, and stability. Looking at the number of PEDOT patents and scientific papers published every year, about 1500 documents per year are produced, highlighting an obvious and remarkable interest for PEDOT from the scientific community. Additionally, about 40% of these figures represent patent applications, demonstrating an additional intense industrial interest in this material.

Over the years, many synthetic strategies have been employed in the development of well-defined oligothiophenes, most of them based on fundamental polymer chemistry approaches [13]. Two main relevant mechanisms for polymerisation are commonly accepted: step-growth and chain-growth methods. These methods provide polymers with distinctly different structures in terms of repeat unit functionality, molecular weight, and dispersity. In particular, chain-growth polymerizations to form addition polymers are most often accomplished using monomers with multiple bonds and loss of unsaturation. In particular, in the chain-growth mechanism, a reactive intermediate is first created in an initiation step and subsequently propagates via repeated monomer addition to provide a macromolecule. When the reactive intermediate is ionic, impurity termination or quenching processes can kill the reactive intermediate, while in the case of radical polymerisation, coupling termination can lead to an overall doubling of the average molecular weight [14]. This approach has been widely used to produce PEDOT. Despite its huge potentialities PEDOT itself does not present high conductivity values; as matter of fact, the conjugated bond structure in PEDOT leads to a rigid conformation, maintaining electron orbital overlap along the backbone of the polymer. Crystallisation is favoured and prevents the material from easily dissolving or melting. As a result, PEDOT has good chemical and thermal stability that makes it difficult to process into thin-film form [15].

Alternatively, a conjugated polymer can be deposited by *in-situ* oxidative polymerisation directly at the surface. On conducting substrates this can be achieved by electrochemical polymerisation, which generally gives coatings of high quality. This method is however rarely suited for large-scale applications and cannot be used to add electronic functionalities to non-conducting surfaces. Furthermore, electropolymerization of EDOT in aqueous media is limited by the low solubility of the monomer. However, it is reported that the addition of surfactants improves its solubility. So it is often required to derivatise the polymer with soluble side chains or to dope the polymer with polyelectrolytes acting as solubilisers in order to be able to process these polymers [16, 17].

According to the latter proposed solution, a particularly conductive form of PEDOT, synthesised with the addition of poly(styrene sulfonic acid) (PSS) and also stabilised by, has been put in place and introduced on the market as an aqueous dispersion under the brandname Baytron, and most recently changed to Clevios [18]. The additional oxidising the neutral polymer prompts conductivity behaviour. The positive charges of PEDOT backbone induced by oxidation are balanced by anions present on PSS. The appropriate choice of the counter-ion highly contributes to the properties of the conductive complex. Pristine PEDOT:PSS yields a conductivity below 0.1 S/cm, too low to be used as an electrode in an efficient solar cell [19]. Despite the appealing properties of this complex, degradation phenomena of PEDOT:PSS have been detected inducing to irreversible morphological and chemical changes which affect film properties [20–22]. As example, an insulating layer of PSS preferentially segregates to the surface of PEDOT:PSS films as identified by Greczynski et al. [23]. The presence of this phase segregation has been also supported by Jukes et al. using small-angle neutron scattering [24]. Given the vertical phase separation, it is not surprising that PEDOT:PSS films exhibit anisotropic conductivities. Conduction, as expected, is inefficient through the depth of the film and enhanced in the plane of the film (10^{-5} and 10^{-3} S/cm, respectively) [25]. It is well known that addition of solvents, such as ethylene glycol, glycerol, dimethyl sulfoxide, and sorbitol significantly improved the conductivity of PEDOT:PSS by up to 2 or 3 orders of magnitude. On the other hand the fabrication processes that involve solvents are often restricted by the solvent-substrate incompatibility as in the case of printed electronics where high importance is played by the interactions between PEDOT:PSS and flexible substrates, mainly consisted of polymers and papers. The non-evaporated solvent could also prompt swelling or degradation phenomena into the non-sensitive substrate after drying step [26].

The synthesis of PEDOT can be also obtained by chemical vapour deposition (CVD) in order to circumvent most of all the disadvantages of liquid phase chemistry approach. CVD is not a new process. Its first practical use was developed in the 1880s in the production of incandescent lamps to improve the strength of filaments by coating them with carbon or metal [27]. CVD may be defined as the deposition of a solid on a heated surface from a chemical reaction in the vapour phase. CVD has several important advantages over the liquid phase chemistry which make it the preferred process in many cases: absence of solvent, high deposition rate, good surface conformity. CVD equipment does not normally require ultrahigh vacuum and generally can be adapted to many process variations. Its flexibility is such that it allows many changes in composition during deposition and the co-deposition of elements or compounds is readily achieved. In this scenario, several kinds of CVD techniques are currently used to obtained PEDOT films.

The first one is the so called plasma-assisted or plasma-enhanced CVD (PE-CVD), introduced in the fabrication of semiconductor from 1960s and for long time restricted to the synthesis of inorganic compounds, becoming an essential factor in the manufacture of semiconductors and other electronic components. This technique is based on electrical energy rather than thermal energy to initiate homogeneous reactions for the production of chemically active ions and radicals that can participate in heterogeneous reactions, which, in turn, lead to layer formation on the substrate. A major advantage of PE-CVD over thermal CVD processes is that deposition can occur at very low temperatures, even close to ambient, which allows the use of temperature sensitive substrates [28]. Unfortunately, monomers injected into the glow discharge are often irregular and fragmentized by the electrons, ions/radicals present into the glow discharge. Furthermore such plasma polymers are often irregular and have rather short chain lengths. Plasma polymerisation leads to a random poly-recombination of radicals and fragments of monomers. A variation could be represented by pulsed-PE-CVD. In this case, the glow discharge is pulsed and when it is switched on (plasma-on), the activate molecules can produce radicals which initiate the polymerisation mechanism. When the glow discharge is switched off (plasma-off), the residual radicals initiate a conventional chemical chain-growth reaction. In this case a more 'conventional' product is obtained but the chain propagation is restricted by the low probability of attaching a new molecule monomer to the radical at the growing chain-end. Moreover the deactivation of chain propagation is expected under the common vacuum conditions (0.1 Torr) owing to the recombination of neighbouring radicals. This leads to a significant loss of the active radical sites during the plasma-off period. Glow discharge is thus re-ignited to reproduce fresh initial radicals. The resulting polymer film has a structure and composition closer to its counterpart produced by liquid phase but it shows repeatable irregularities induced by the plasma-on period [29].

Another known CVD method is the vapour phase polymerisation (VPP), where the solubility of the conjugated monomer is no longer required. VPP directly translates the step-growth mechanism established for the liquid phase synthesis via solvent-less environment [30]. VPP is usually performed by two-step process. The first one consists of the pre-application of the oxidant to the substrate, which is typically carried out by spin coating, while the second step consists of the exposure of the pre-treated substrate to monomer vapours under vacuum reaction chamber. The VPP delivers conducting polymers, like polypyrrole and PEDOT, using iron(III) chloride as oxidant [31, 32]. Spreading the oxidant all over the surface limits the use of VPP to non-liquid sensitive substrate [33]. Analogous to VPP, another process called oxidative CVD (oCVD) has been developed by Prof. K.K. Gleason at Massachusetts Institute of Technology [34]. The oCVD is a single step process, where both oxidant and monomer, in vapour phase, are delivered to the substrate into the chamber reaction. This is the main difference between the VPP and oCVD methods. Deposition of PEDOT films by oCVD using iron(III) chloride has been described in detail in the literature [35]. With this method smooth PEDOT films with conductivity higher than 1000 S/cm have been obtained [36]. Actually one of main drawback of using iron(III) chloride, as oxidant compound in the polymerisation of EDOT, lies in the use of post processing step carried out with methanol to remove the unreacted iron (III) chloride and by-products as iron(II) chloride as well as any oligomers or short chains formed during the polymerisation process as they become contaminant during device fabrication. Additionally it is also rather difficult to provide sufficient oxidant flux since it is often a powder [37].

Halogens have been widely used as a dopant for making conducting polymers [38, 39]. So in an attempt to make oCVD process completely dry, bromine has been used as the replacement for iron(III) chloride. Use of bromine resulted in a completely dry process with no post processing required. The achieved conductivity was around 380 S/cm but details on surface area covered by PEDOT film as well as resistance over the time were not disclosed. Despite these encouraging results any further papers using bromine as oxidant have been published [40]. In this framework, quite recently a novel CVD technique has been developed by Dr. D. Lenoble and co-workers at Luxembourg Institute of Science and Technology (LIST). The process, named Plasma Radicals Assisted Polymerisation via CVD (PRAP-CVD), has been demonstrated as an efficient alternative to CVD processes in depositing conductive thin films [41]. This method is directed to form conjugated polymer directly on a substrate.

2. Experimental

PEDOT has been investigated as case study to prove the effectiveness of PRAP-CVD.

PRAP-CVD polymerisation results from surface reaction of the monomer and oxidant flows, here 3,4-ethylenedioxythiophene (EDOT) and bromine, respectively [41]. The oxidation of EDOT to form PEDOT shows some similarities to the oxidative polymerisation mechanism of pyrrole suggested by Diaz [42, 43]. The proposed polymerisation mechanism of EDOT to form PEDOT is illustrated in **Figure 1** [41].

A preliminary step takes place in a plasma chamber, where bromine radicals are generated while the glow discharge is switched on. Progressively they reach the reaction chamber where EDOT monomer is parallel injected. Bromination of EDOT molecules occurred in α position by the formation of a radical that shows several resonance forms. A dimer is formed by combination of two of these radicals. Substitution of the EDOT thiophene ring at the 3,4-positions blocks β-coupling, allowing new bonds only at the 2,5-positions. The dimer can be oxidised to form another radical following the steps described above. The alternating single and double bonds of the oligomers are π-conjugated, which delocalizes the electrons

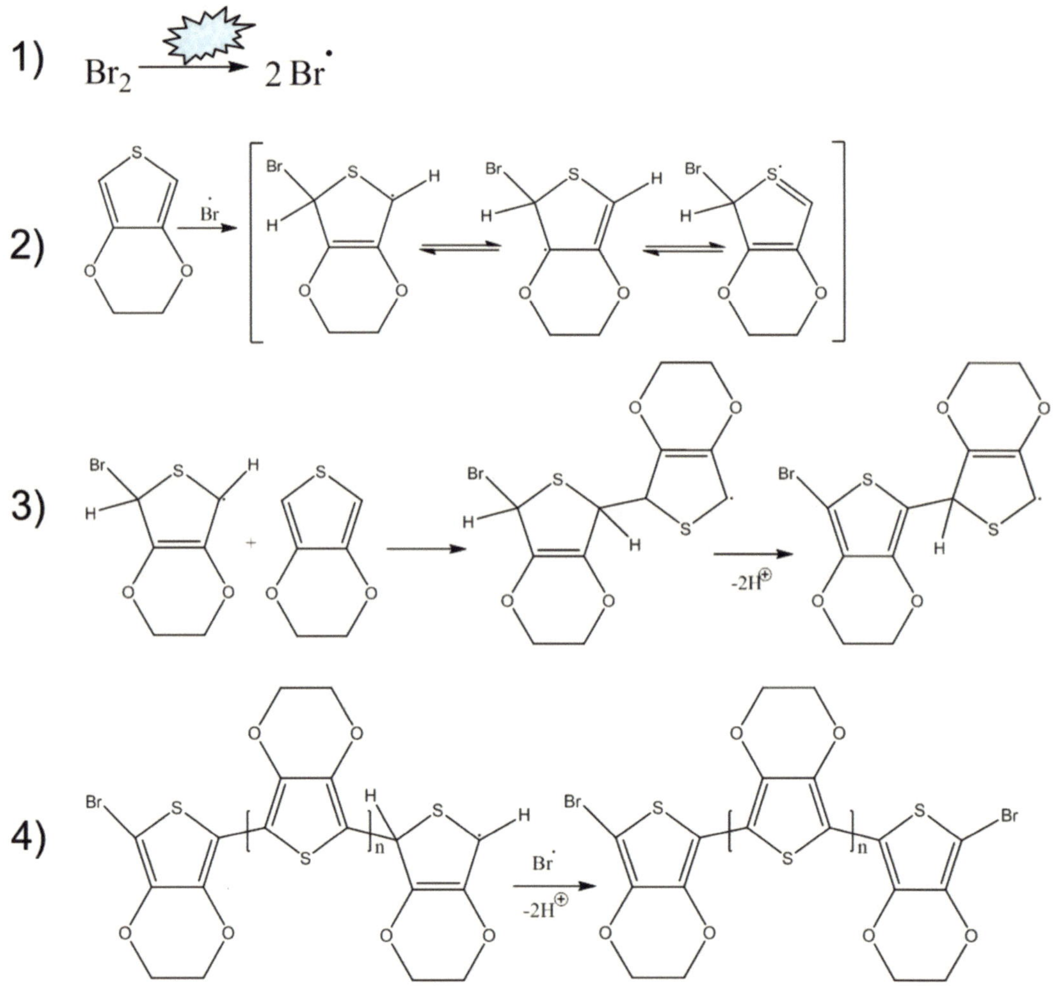

Figure 1. Polymerisation mechanism of EDOT to form PEDOT. (1) Ignition of glow discharge fed with bromine gas allows the production of bromine radicals. (2) Bromination of EDOT molecules occurs in α position. (3) Radical polymerisation propagates. (4) Polymerisation stops when also α' position is brominated [41].

and decreases the oxidation potential [34]. In PRAP-CVD the bromine acts as radical initiator as well as oxidant, and the final electronic arrangement presented in the polymer is depicted in **Figure 2**.

PRAP-CVD PEDOT, which is carried out in PRODOS-200 PVPD™ R&D System (AIXTRON SE, Herzogenrath, Germany), owing to the facilities of the machine and the mild experimental conditions, can be deposited on a wide range of substrates, both in terms of dimension and in terms of composition: from silicon wafer to fabrics and up to 15 × 15 cm^2, as reported in **Figure 3**.

Raman studies were conducted to determine the vibrational modes of the different regions in the wavenumber range of 250–1800 cm^{-1}, and to prove the obtained EDOT polymerisation.

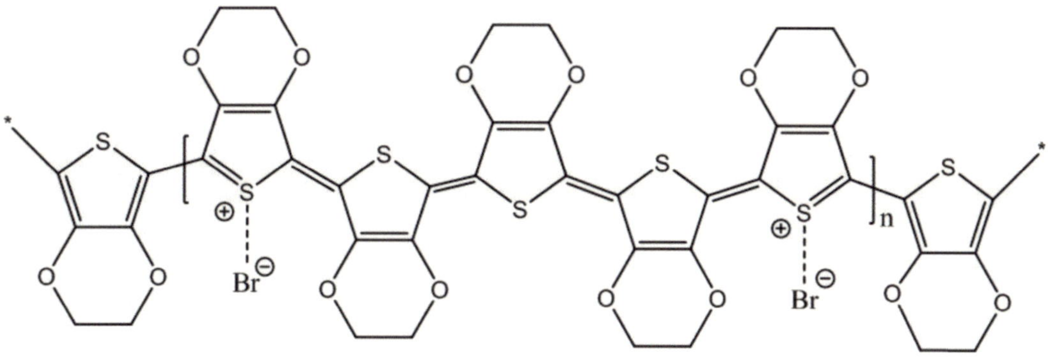

Figure 2. Electronic arrangement of Br-PEDOT films. The bromine acts as radical initiator as well as oxidant [41].

Figure 3. Examples of substrates treated via PRAP-CVD to obtain PEDOT films: (a) silicon wafer (20 cm in dia), (b) glass (15 × 15 cm^2), and (c) PET fabric (12 × 12 cm^2) [41, 44].

PEDOT was deposited on silicon wafer, prevented covered with a thin layer of SiO_2 (deposited in house by Rapid Thermal-Chemical Vapour Deposition – RT-CVD) to avoid any interferences during the electrical measurements, as reported in **Figure 4**.

The spectrum of monomer was compared with that one of polymer. Six strong Raman bands were detected in the spectrum of EDOT, **Figure 4a**, at 1486, 1423, 1185, 892, 834, and 766 cm^{-1} [45]. The shape of spectra was completely different and the spectrum of PEDOT showed much more bands related to different vibration modes owing to presence of the polymer chain. In **Table 1** the assignment related to the bands are reported. A good agreement with literature data was found to support the accuracy of the proposed polymerisation mechanism [46, 47].

Furthermore, the confirmation that the polymerisation took place in the right way, i.e. without thiophene opening ring, in favour of correct PEDOT film formation, came also from the absence of the band at 1705 cm^{-1} [48].

The mild experimental conditions carried out in the PRAP-CVD allow even the treatment of sensible substrates as papers of fabrics without any damages. To prove this advantage of PRAP-CVD compared to the other CVDs, a few temperature sensitive substrates have been treated and a PEDOT thin layer was deposited. SEM micrographies were recorded and shown in **Figure 5** where (a) PET fabric, (b) cotton fabric, (c) linen fabric and (d) paper cellulose fibres are reported.

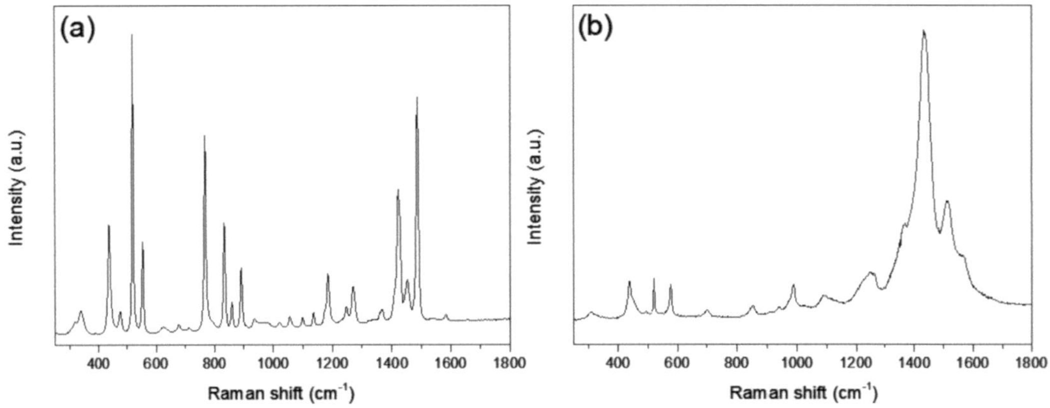

Figure 4. Raman spectrum of (a) EDOT and (b) PRAP-CVD PEDOT films recorded at an excitation wavelength of 633 nm.

Wavenumber (cm^{-1})	Assignments
1509	Asymmetric $\nu C\alpha = C\beta$
1432	Symmetric $\nu C\alpha = C\beta$ (–O)
1366	$\nu C\beta$-$C\beta$
1266	Inter-ring $\nu C\alpha$-$C\alpha$
1096	νC-O-C
990, 573	Oxyethylene ring deformation
699	Symmetric νC-S-C
439	δSO_2

Table 1. Observed frequencies of doped poly(3,4-ethylenedioxythiophene) with assignment of principal bands.

Samples have been investigated after the deposition without any further manipulation. Not only by PRAP-CVD it is possible to treat fibres but PRAP-CVD benefits of a quite unique property in terms of conformal deposition on complex substrates, as of the matter of fact, fibres are, **Figure 6**.

PET fabrics have been chosen as reference substrate for this series of characterizations. Correspondence of the textile morphology between the uncoated and coated sample illustrated the high conformal coverage of the fibres with polymer thin film, **Figure 6a** and **b**. The presence of PEDOT was confirmed by Energy Dispersive X-ray (EDX) spectroscopy investigation reported in **Figure 6c** and **d**, respectively. In the spectrum of PET, peaks of carbon (C) Ka at 0.3 keV and oxygen (O) at 0.5 keV were detected while the spectrum of fibres treated via PRAP-CVD highlighted the presence of sulphur (S) at 2.3 keV and bromine (Br) at 11.9 keV. Sulphur as well as carbon and oxygen could be attributed to the chemical composition of EDOT while bromine to the dopant. PEDOT film resulted in defect/pin-hole free, as attested by the high magnification SEM micrographs. In order to deeply prove this

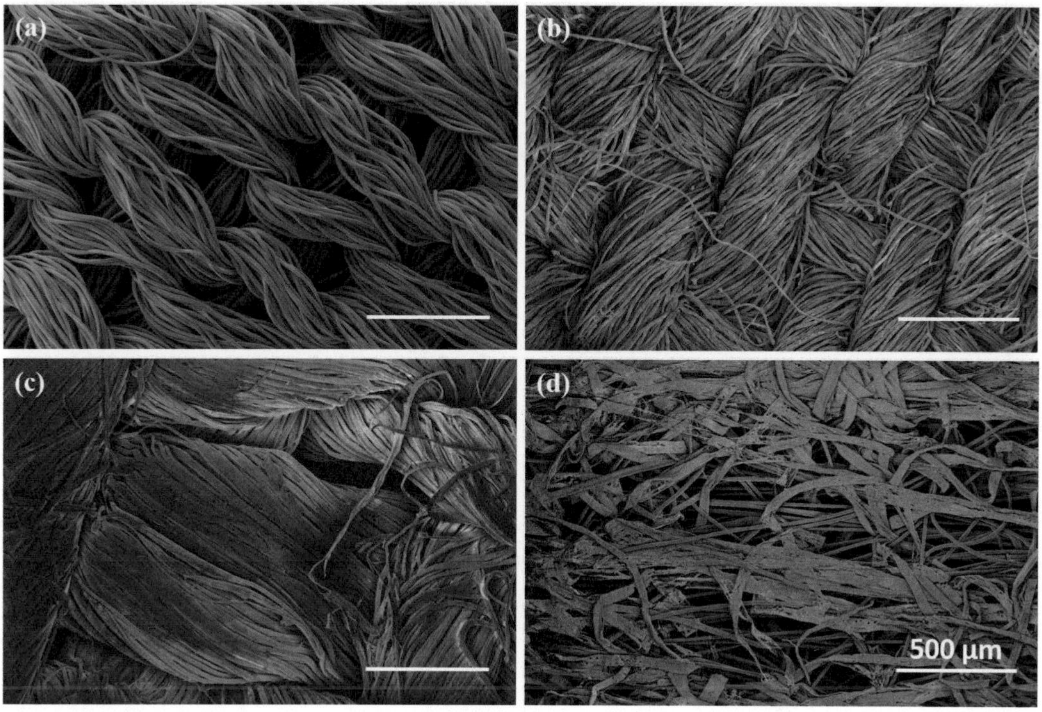

Figure 5. SEM microscopies of PRAP-CVD PEDOT film deposited on (a) PET fabric, (b) cotton fabric, (c) linen fabric and (d) paper cellulose fibres. Samples have been investigated after the deposition without any further manipulation [44].

peculiarity of PRAP-CVD, a stripe of PET fabric covered with PEDOT was cut by means of a scalpel and analysed. In **Figure 7**, micrographs of the same PEDOT/PET area, recorded at different magnifications, are reported (**Figure 7a** and **c**).

The cut induced a physical convergence of fibres. The lighter part of image could be attributed to the PEDOT which responded differently from PET under the irradiation of the electron beam. In the same experimental conditions, PET, as all non-conductive materials, suffers of charging effects, revealing a darker image than PEDOT/PET. The same images were also artificially coloured by Image J software as reported in **Figure 5b** and **d** to easily identify the layer of PEDOT which conformably surrounded each fibre although the convergence of fibres. By SEM it was possible on one side to investigate the conformality of the film and on the other anticipated its electrical conductivity properties too. Following FIB cross-sectioning step, an accurate thickness measurement of film could be determined. A thickness of (215 ± 10) nm allowed to establish a deposition rate of 6 nm/min. This level of coating quality is unequalled to current wet and dry PEDOT synthesis. PRAP-CVD allows to modify only the surface of substrate without impacting on the bulk properties, as demonstrated by XPS spectra in **Figure 8** [41].

Chemical states of carbon (C 1s) in PET and PEDOT/PET C 1s were identified by XPS. In **Figure 8a** three chemical states of C 1s, reflecting the presence of the three peaks, were

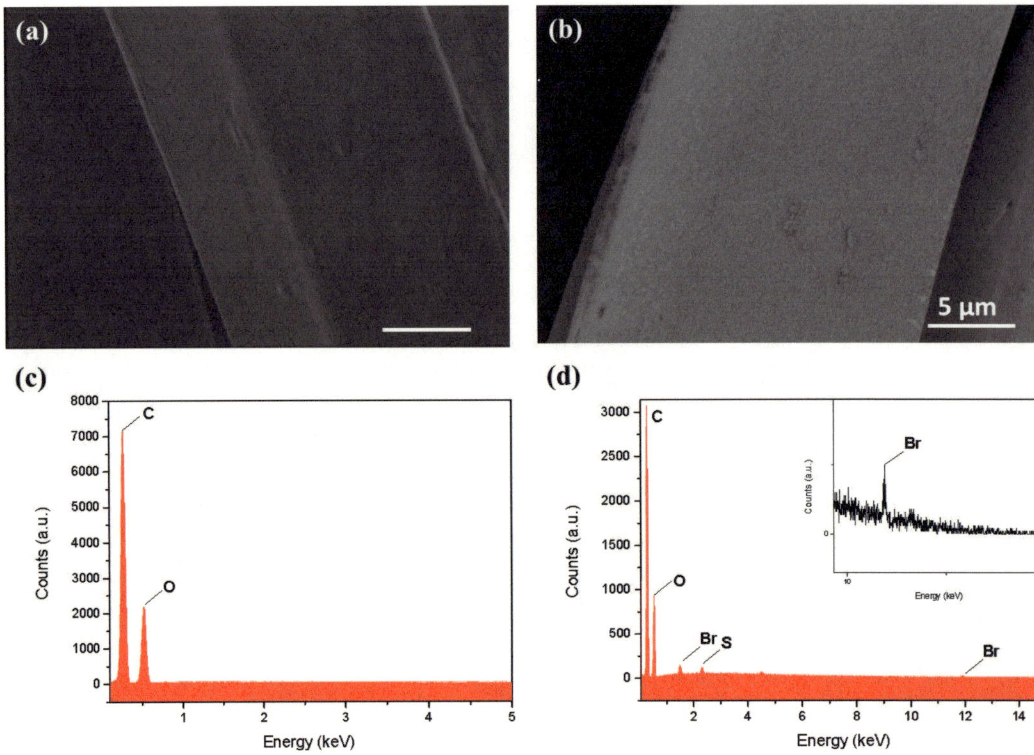

Figure 6. SEM images of textile fibres (a) before and (b) after the PRAP-CVD process. Owing to the induced conductivity of PEDOT, the two images resulted in different colours, darker the raw PET and lighter the PEDOT/PET fibre. Respective EDX spectra identified the presence of (c) carbon and oxygen in raw PET and (d) also sulphur and bromine residues in the PEDOT coated fabric [44].

identified in the PET. They could be attributed to C–C/H bond (aliphatic/aromatic carbon atoms) (C1 - BE: 284.5 eV), C–O bond (methylene carbon atoms singly bonded to oxygen) – (C2 – BE: 286.0 eV), –COO– bond (ester carbon atoms) (C3 – BE: 288.5 eV) and π–π^* shake-up transition associated with the aromatic ring at 291 eV. By performing the best fitting procedure, the following percentage areas: 54.2% C1, 24.6% C2, 14.6% C3, and 6.6% shake-up were obtained [49]. In **Figure 8b**, the distribution of functional groups of PEDOT/PET C 1s showed three different components: C–C/H bonds (C1 – BE: 284.4 eV), C–S bonds (C2 – BE: 285.5 eV) and C–O bonds (C3 – BE: 286.1 eV). Additionally, an asymmetrical peak at about 288 eV was also identified and attributed to the contribution from the π–π^* shake-up transition and positively polarised or charged carbon [50]. Peak areas were calculated as 34.8% for C–C/H, 42.5% for C–S, 17.3% for C–O and 5.4% for the shake-up, respectively [51]. Best-fitting procedure was also applied to S 2p core-level, which corresponded to single sulphur bonding environment in PEDOT with a spin-split doublet, 2 p1/2 and 2 p3/2 separated by 1.18 eV in binding energy and an area ratio 1:2, as shown in **Figure 9**.

The pick presented an asymmetric tail at higher binding energy, which could be related to the doping process, where the delocalized π-electrons in thiophene ring broaden the binding

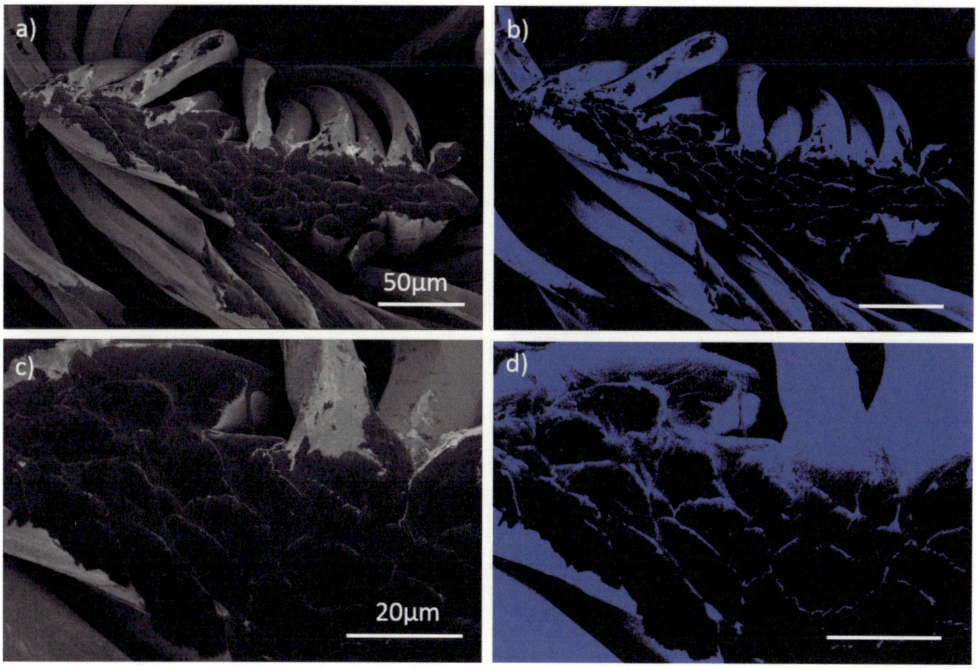

Figure 7. SEM micrographs of PET fibres cross-section with PEDOT film (a, c) at different magnification; (b, d) corresponded coloured picture at same magnification. The darker area corresponded to PET fibres. Conductive PEDOT film responded differently from PET while irradiated by the electron beam [44].

energy spectrum of the sulphur atom [52]. The quantitative analysis of PEDOT film showed the followed atomic percentage of 66% C, 21% O, 9% S, and 4% Br [41].

In the last years the demand of transparent flexible and conductive electrodes increased suddenly, in particular the research of materials to replace indium tin oxide (ITO). The main drawbacks presented by ITO are on one side the cost of indium itself and the remaining amount of this element the earth and on the other the brittleness of ITO based electrodes. PRAP-CVD PEDOT could answer to these requirements because as demonstrated it can be deposited on flexible substrate as plastic and it shows optical transparency properties, as reported in **Figure 3**. More in detail, transmittance and absorbance values of PEDOT deposited on glass in UV range were recorded and presented in **Figure 10**.

In **Figure 10a**, the transmittance of PEDOT thin films decreased while wavelength increased beyond 500 nm, this is owing to the presence of the free carrier tail but PEDOT still showed a value of transmittance over 70%.

In **Figure 10b**, the peculiar absorption feature, which characterises PEDOT film, known in the literature as a 'free carrier tail', is confirmed and can be attributed to the conductivity of the polymer films [53]. The presence of this band has been found and investigated for the first time in the doped polyaniline. It corresponds to the polymer having a longer conjugation length and greater order, which allows for greater mobility of charge carriers [54].

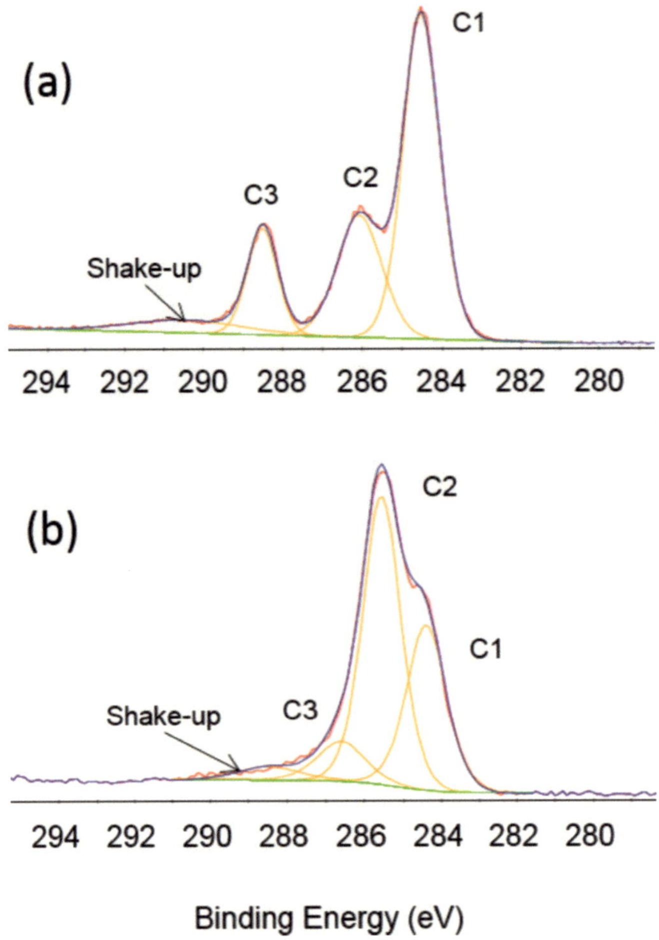

Figure 8. XPS C 1s high resolution spectra of (a) PET and (b) PEDOT/PET and relatively chemical states [44].

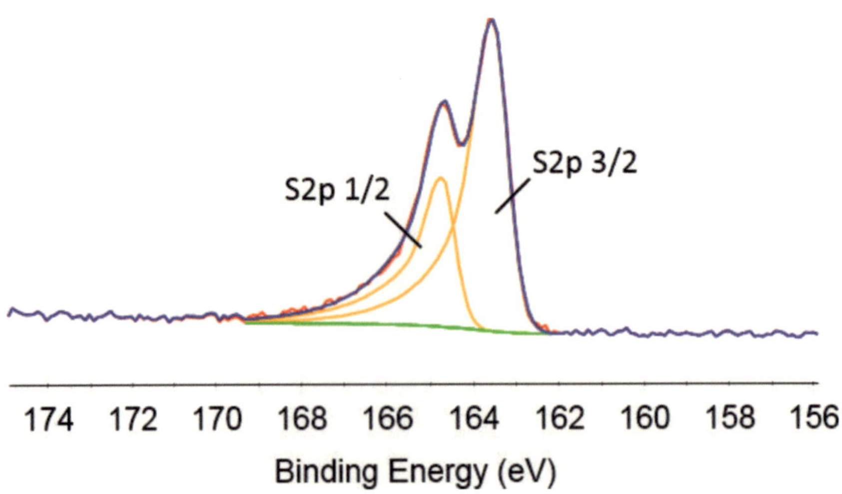

Figure 9. XPS HR high resolution spectra of S 2p.

http://dx.doi.org/10.5772/intechopen.71736

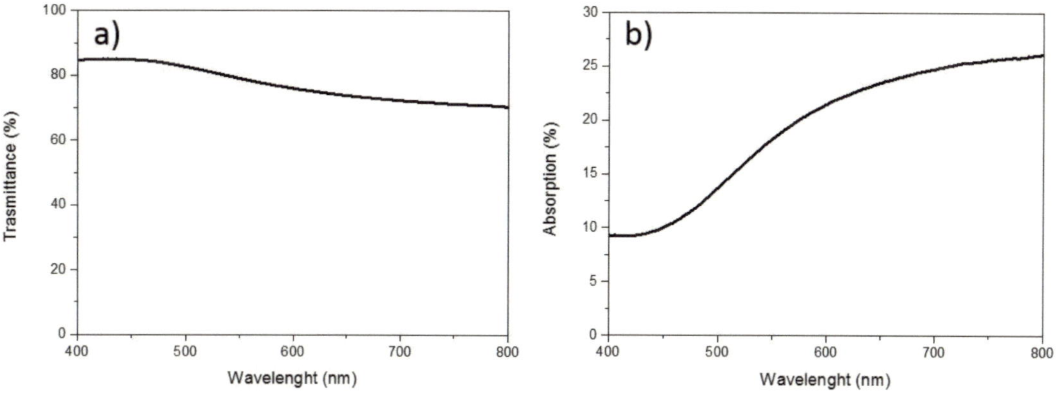

Figure 10. UV-Vis-NIR spectrum of thin PEDOT films as a function of (a) transmittance and (b) absorbance.

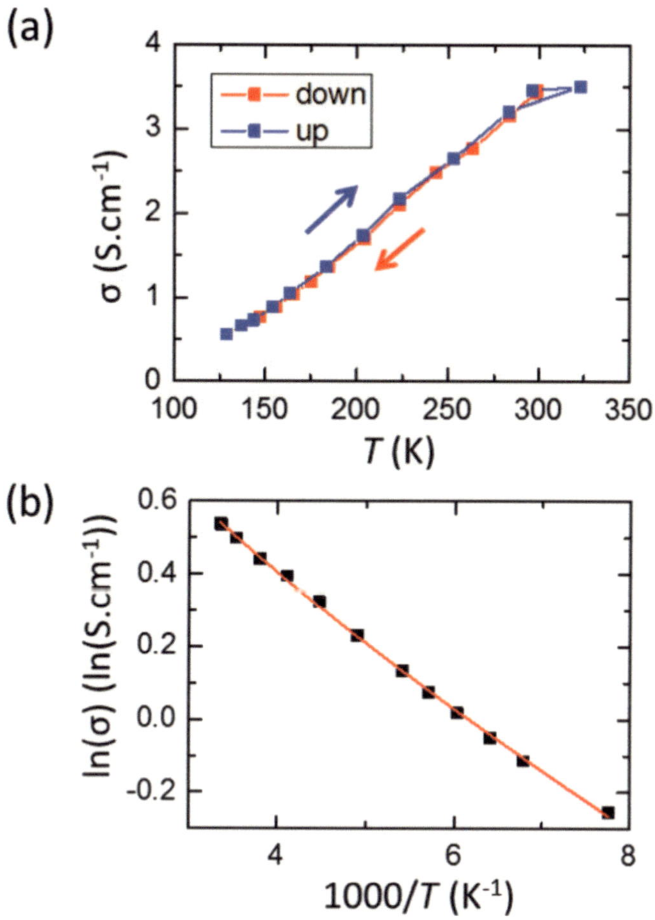

Figure 11. (a) Conductivity of the 50 nm thick PEDOT film deposited on BOROFLAT® 33 as a function of the sample temperature. The measurements were performed both during cooling-down (red) and warming-up (blue) the sample. (b) Natural logarithm of the conductivity plotted as a function of 1000/T (black points) for the data measured during the cooling-down. The red curve is a fit of the data by Eq. (1), with the power $\alpha = 0.63 \pm 0.07$.

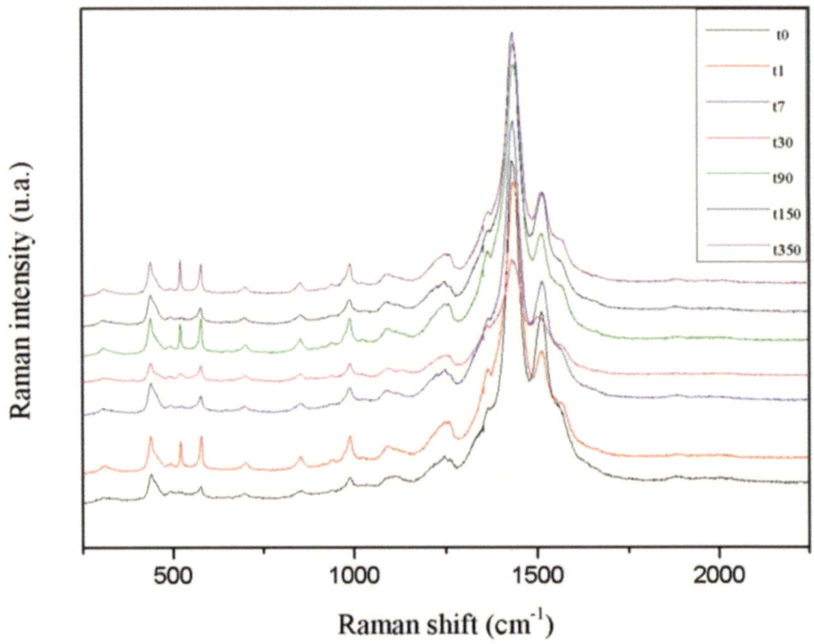

Figure 12. Raman spectra of PEDOT film recorded all along 1 year [41].

Performances of films have been investigated also in terms of conductivity as function of temperature. **Figure 11** shows the conductivity as a function temperature in the range 130–350 K, both when cooling-down and warming-up the sample.

The good reproducibility showed that there was no structural changes in the film, and that the temperature-dependence of the conductivity was owing to the intrinsic carrier transport mechanism in the film. The observed decreasing in conductivity, while the temperature decreases, is typical for an organic conductor, where carrier transport usually occurs by hopping conduction. In order to further investigate the carrier transport mechanism, in **Figure 11b** the natural logarithm of the conductivity as a function of the inverse temperature, 1/T have been reported. It was observed a good fit with a power, implying the following dependence for the conductivity as reported in Eq. (1):

$$\sigma = \sigma_0 \exp\left[-\left(\frac{T_0}{T}\right)\alpha\right] \quad (1)$$

The best fit was obtained with $\alpha = 0.63 \pm 0.07$. This value is far from 1/4, already reported for 3D variable range hoping (VRH), far from 1, as reported for nearest neighbour hoping, and far from 1/3, as reported for 2D variable range hoping [25, 55, 56]. The value is closer to 1/2, already reported for PEDOT:PSS and which corresponds to 1D VRH or hopping in a granular material where, the exponent α for 3D VRH varies from1/4 to 1/2 due to charging energy-limited tunnelling between grains [57, 58].

Another requirement that PEDOT film has to show is the stability of the time. PEDOT films were exposed to ambient conditions over 1 year and its chemical structure was followed by Raman spectroscopy investigation, as reported in **Figure 12**.

Previous studies suggested that exposure to water vapour in the atmosphere could induce an increase in PEDOT, in particular PEDOT:PSS, sheet resistance as the hygroscopic polymer absorbs moisture over time [59, 60]. As shown in **Figure 12**, the main Raman peaks did not suffer of changing during the year. This different behaviour can be explained by the absence of PSS, which reveals all its acid character when exposed to the moisture.

3. Conclusions

PRAP-CVD has been successfully demonstrated to be a novel promising technique to deposit conjugated polymers. In particular, PRAP-CVD of EDOT leading to the formation of uniform conducting doped PEDOT film, by using bromine as oxidant, has been attested. Raman spectroscopy confirmed the chemical composition of PRAP-CVD PEDOT films. The use of bromine, as oxidant, allows to avoid every supplementary rinsing process, making the process completely dry, comparing to the PSS. All issues related to the corrosion, owing to the acidic behaviour of PSS, are bypassed in the PRAP-CVD. The use of bromine and the absence of PSS allow the use of PRAP-CVD on a wide range of substrates which could be damaged by a wet step like papers, plastics, etc. as demonstrated above. 3D samples benefit of a high conformal coverage. PRAP-CVD, developed in a β-side machine version, allows the treatment of quite large samples compared to VPP. This technique can be easily scaled up featuring highly competitive growth rate compared to other vapour phase growth of PEDOT thin films. Finally PRAP-CVD PEDOT film is transparent in the visible range and stable over the year.

Acknowledgements

The authors wish to express their appreciation to Mr. Noureddine Adjeroud for his contribution to RT-CVD process; to Dr. Mael Guennou, Dr. Jérôme Guillot, for performed Raman and XPS analyses respectively. Aixtron SE is also acknowledge for direct technical assistance.

This work has been funded in part by Fond National de la Recherche Luxembourg (FNR Luxembourg) through INTER/NSF/MAT/11/01 'Visible Light Nanocomposite Photocatalysts (VISICAT) project, and in part by AFR call PDR 2012-2, ref 4735153.

Appendices and nomenclatures

BE	Bonding Energy
CVD	Chemical vapour deposition
EDX	Energy Dispersive X-ray
ITO	Indium tin oxide

LIST	Luxembourg Institute of Science and Technology
OFET	Organic field-effect transistors
OLED	Organic light-emitting diode
OPV	Organic photovoltaics
oCVD	Oxidative chemical vapour deposition
PE-CVD	Plasma enhanced-chemical vapour deposition
PRAP-CVD	Plasma Radicals Assisted Polymerisation via CVD
PEDOT	Poly(3,4-ethylenedioxythiophene)
PSS	Poly(styrene sulfonic acid)
PVPD	Polymer vapour phase deposition
RT-CVD	Rapid Thermal-Chemical Vapour Deposition
VPP	Vapour phase polymerisation

Author details

Bianca Rita Pistillo*, Kevin Menguelti, Didier Arl, Renaud Leturcq and Damien Lenoble

*Address all correspondence to: biancarita.pistillo@list.lu

Luxembourg Institute of Science and Technology, Belvaux, Luxembourg

References

[1] Dall'Olio A, Dascola G, Varacca V, Bocche V. Résonance paramagnétique électronique et conductivité d'un noir d'oxypyrrol électrolytique. Comptes Rendus de l'Académie des Sciences. 1968;**C267**:433

[2] Letheby H. On the production of a blue substance by the electrolysis of sulphate of aniline. Journal of the Chemical Society. 1862;**15**:161-163

[3] Shirakawa H. Nobel Lecture: The discovery of polyacetylene film — The dawning of an era of conducting polymers. Reviews of Modern Physics. 2001;**73**:713-718

[4] MacDiarmid AG. Nobel Lecture: "Synthetic metals": A novel role for organic polymers. Reviews of Modern Physics. 2001;**73**:701-712

[5] Heeger AJ. Nobel Lecture: Semiconducting and metallic polymers: Generation of polymeric materials. Reviews of Modern Physics. 2001;**73**:681-700

[6] Chiang CK, Fincher CR, Park JYW, Heeger AJ, Shirakawa H, Louis EJ, et al. Electrical conductivity in doped polyacetylene. Physical Review Letters. 1977;**39**:1098

[7] Yamamoto T, Sanechika K, Yamamoto A. Preparation of thermostable and electric-conducting poly(2,5-thienylene). Journal of Polymer Science Part C: Polymer Letters Edition [Internet]. 1980;**18**(1):9-12. Available from: http://onlinelibrary.wiley.com/doi/10.1002/pol.1980.130180103/full

[8] Lin JW-P, Dudek LP. Synthesis and properties of poly(2,5-thienylene). Journal of Polymer Science, Part A: Polymer Chemistry Edition [Internet]. 2002;**18**(9):2869-2873. Available from: http://doi.wiley.com/10.1002/pol.1980.170180910%5Cnpapers3://publication/uuid/FB763B27-2323-49FF-A881-E553F325B648%5Cnhttp://doi.wiley.com/10.1002/pol.1980.170180910%5Cnpapers3://publication/doi/10.1002/pol.1980.170180910

[9] Sirringhaus H, Bird M, Richards T, Zhao N. Charge transport physics of conjugated polymer field-effect transistors. Advanced Materials. 2010;**22**(34):3893-3898

[10] Laquai F, Park Y-S, Kim J-J, Basché T. Excitation energy transfer in organic materials: From fundamentals to optoelectronic devices. Macromolecular Rapid Communications [Internet]. 2009;**30**(14):1203-1231. Available from: http://doi.wiley.com/10.1002/marc.200900309

[11] Weickert J, Dunbar RB, Hesse HC, Wiedemann W, Schmidt-Mende L. Nanostructured organic and hybrid solar cells. Advanced Materials. 2011;**23**(16):1810-1828

[12] Jonas F, Krafft W. (Bayer AG) European Patent EP 440 957, March 10, 1991

[13] Henssler JT, Zhang X, Matzger AJ. Thiophene/thieno[3,2-b]thiophene co-oligomers: Fused-ring analogues of sexithiophene. The Journal of Organic Chemistry. 2009;**74**(23):9112-9119

[14] Ternay AL. Contemporary Organic Chemistry. 2nd ed. Philadelphia: Saunders; 1979. 1277 p. ISBN: 0721687954

[15] Brédas JL, Silbey R. Conjugated Polymers: The Novel Science and Technology of Highly Conducting and Nonlinear Optically Active Materials. Dordrecht: Publisher Springer Netherlands; 1991. 624 p. ISBN: 978-94-011-3476-7

[16] Sakmeche N, Aeiyach S, Aaron J, Jouini M, Lacroix JC, Lacaze P. Improvement of the electrosynthesis and physicochemical properties of poly (3, 4-ethylenedioxythiophene) using a sodium dodecyl sulfate micellar aqueous medium. Langmuir. 1999;**1**(28):2566-2574

[17] Vasantha VS, Phani KLN. Effect of hydroxypropyl-β-cyclodextrin on the electrochemical oxidation and polymerization of 3,4-ethylenedioxythiophene. Journal of Electroanalytical Chemistry. 2002;**520**(1-2):79-88

[18] Andreas Elschner, Stephan Kirchmeyer, Wilfried Lovenich, Udo Merker, Knud Reuter. PEDOT: Principles and Applications of an Intrinsically Conductive Polymer. Broken Sound Parkway NW: CRC Press, Taylor & Francis Group; 2010. 377 p. ISBN: 9781420069112

[19] Groenendaal L, Jonas F, Freitag D, Pielartzik H, Reynolds JR. Poly(3,4-ethylenedioxythiophene) and its derivatives: Past, present, and future. Advanced Materials. 2000;**12**(7):481-494

[20] Chung J, Kim KH, Lee JC, Kim MK, Shin HJ. Spectromicroscopic investigation of polymer light-emitting device degradation. Organic Electronics: Physics, Materials, Applications. 2008;**9**(5):869-872

[21] Fehse K, Meerheim R, Walzer K, Leo K, Lövenich W, Elschner A. Lifetime of organic light emitting diodes on polymer anodes. Applied Physics Letters. 2008;**93**(8):2006-2009

[22] Martin SJ, Jones RAL, Geoghegan M, Higgins AM, Grizzi I, Halls JJM, et al. Current-induced chain migration in semiconductor polymer blends. Physical Review b - Condensed Matter and Materials Physics. 2005;**71**(8):1-4

[23] Greczynski G, Kugler T, Salaneck WR. Characterization of the PEDOT-PSS system by means of X-ray and ultraviolet photoelectron spectroscopy. Thin Solid Films. 1999;**354**(1):129-135

[24] Jukes PC, Martin SJ, Higgins AM, Geoghegan M, Jones RAL, Langridge S, et al. Controlling the surface composition of poly(3,4-ethylene dioxythiophene)–poly(styrene sulfonate) blends by heat treatment. Advanced Materials [Internet]. 2004;**16**(9):807-811. Available from: http://doi.wiley.com/10.1002/adma.200306487

[25] Nardes AM, Kemerink M, Janssen RAJ. Anisotropic hopping conduction in spin-coated PEDOT:PSS thin films. Physical Review B [Internet]. 2007;**76**(8):085208(1-7). Available from: http://link.aps.org/doi/10.1103/PhysRevB.76.085208

[26] Angiolini S, Avidano M, Bracco R, Barlocco C, Young NG, Trainor M, et al. High performance plastic substrates for active matrix flexible FPD. In: High Performance Plastic Substrates for Active Matrix Flexible FPD. John Wiley & Sons; 2003. pp. 1325-1327

[27] Pierson HO. Powell CF, Oxley JH, Blocher JM Jr, editors. Handbook of Chemical Vapor Deposition. New York: John Wilewy & Sons; 1966. 435

[28] Favia P, Sardella E, Lopez LC, Laera S, Milella A, Pistillo BR, et al. Plasma Assisted Surface Modification Processes for Biomedical Materials and Devices. In: Güçeri S, Fridman A, editors. Plasma Assisted Decontamination of Biological and Chemical Agents [Internet]. Cesme-Izmir, Turkey: Springer; 2008. pp. 203-225. Available from: http://dx.doi.org/10.1007/%3E978-1-4020-8439-3_17

[29] Zhu Z, Liu C, Xu J, Jiang Q, Shi H, Liu E. Improving the electrical conductivity of PEDOT:PSS films by binary secondary doping. Electronic Materials Letters. 2016;**12**(1): 54-58

[30] Mohammadi A, Hasan MA, Liedberg B, Lundström I, Salaneck WR. Chemical vapour deposition (CVD) of conducting polymers: Polypyrrole. Synthetic Metals. 1986;**14**(3):189-197

[31] Winther-Jensen B, West K. Vapor-phase polymerization of 3,4-ethylenedioxythiophene: A route to highly conducting polymer surface layers. Macromolecules. 2004;**37**(12): 4538-4543

[32] Abbett KF, Teja AS, Kowalik J, Tolbert L. Iodine doping of poly (3-undecylbithiophene) and its composites with polystyrene using supercritical carbon dioxide. Journal of Applied Polymer Science. 2003;**90**:3876-3881

[33] Thompson BC, Winther-Jensen O, Vongsvivut J, Winther-Jensen B, MacFarlane DR. Conducting polymer enzyme alloys: Electromaterials exhibiting direct electron transfer. Macromolecular Rapid Communications. 2010;**31**(14):1293-1297

[34] Lock JP, Im SG, Gleason KK. Oxidative chemical vapor deposition of electrically conducting poly(3,4-ethylenedioxythiophene) films. Macromolecules. 2006;**39**(16):5326-5329

[35] Im SG, Olivetti EA, Gleason KK. Systematic control of the electrical conductivity of poly (3,4-ethylenedioxythiophene) via oxidative chemical vapor deposition (oCVD). Surface and Coatings Technology. 2007;**201**(22-23 SPEC. ISS.):9406-9412

[36] Baxamusa SH, Im SG, Gleason KK. Initiated and oxidative chemical vapor deposition: A scalable method for conformal and functional polymer films on real substrates. Physical Chemistry Chemical Physics [Internet]. 2009;**11**(26):5227. Available from: http://xlink.rsc.org/?DOI=b900455f

[37] Im SG, Gleason KK. Systematic control of the electrical conductivity of poly(3,4-ethylenedioxythiophene) via oxidative chemical vapor deposition. Macromolecules [Internet]. 2007;**40**(18):6552-6556. Available from: http://pubs.acs.org/doi/abs/10.1021/ma0628477

[38] Mathai CJ, Saravanan S, Anantharaman MR, Venkitachalam S, Jayalekshmi S. Effect of iodine doping on the bandgap of plasma polymerized aniline thin films. Journal of Physics D: Applied Physics. 2002;**35**(17):2206-2210

[39] Gal YS, Junc BAL, Chol SAMK. Electrical conductivity of poly (2-ethynylthiophene) and poly (2-ethynylfuran) doped with electron acceptors. Journal of Applied Polymer Science. 1991;**42**:1793-1797

[40] Chelawat H, Vaddiraju S, Gleason K. Conformal, conducting poly(3,4-ethylenedioxythiophene) thin films deposited using bromine as the oxidant in a completely dry oxidative chemical vapor deposition process. Chemistry of Materials [Internet]. 2010;**22**(9):2864-2868. Available from: http://dx.doi.org/10.1021/cm100092c

[41] Pistillo BR, Menguelti K, Desbenoit N, Arl D, Leturcq R, Ishchenko OM, et al. One step deposition of PEDOT films by plasma radicals assisted polymerization via chemical vapour deposition. Journal of Materials Chemistry C [Internet]. 2016;**4**:5617-5625. Available from: http://xlink.rsc.org/?DOI=C6TC00181E

[42] Sadhir RK, Schoch KF. Preparation and properties of plasma-polymerized thiophene (PPT) conducting films. Thin Solid Films. 1993;**223**(1):154-160

[43] Genies EM, Bidan G, Diaz AF. Spectroelectrochemical study of polypyrrole films. Journal of Electroanalytical Chemistry and Interfacial Electrochemistry. 1983;**149**(1983):101-113

[44] Pistillo BR, Menguelti K, Arl D, Addiego F, Lenoble D. PRAP-CVD: How to design high conformal PEDOT surfaces. RSC Advances [Internet]. 2017;**7**(31):19117-19123. Available from: http://xlink.rsc.org/?DOI=C7RA00343A

[45] Selvaganesh SV, Mathiyarasu J, Phani KLN, Yegnaraman V. Chemical synthesis of PEDOT–Au nanocomposite. Nanoscale Research Letters [Internet]. 2007;**2**(11):546-549. Available from: http://www.nanoscalereslett.com/content/2/11/546

[46] Farah AA, Rutledge SA, Schaarschmidt A, Lai R, Freedman JP, Helmy AS. Conductivity enhancement of poly(3,4-ethylenedioxythiophene)-poly(styrenesulfonate) films post-spincasting. Journal of Applied Physics [Internet]. 2012;**112**(11):113709-1-8. Available from: http://scitation.aip.org/content/aip/journal/jap/112/11/10.1063/1.4768265

[47] Garreau S, Louarn G, Buisson JP, Froyer G, Lefrant S. In situ spectroelectrochemical Raman studies of poly(3,4-ethylenedioxythiophene) (PEDT). Macromolecules. 1999;**32**(20):6807-6812

[48] Robert M Silverstein, Francis X Webstein, David J Kiemle. Spectrometric Identification of Organic Compounds. Organic Chemistry; John Wiley & Sons. 2005. pp. 1-550

[49] Beamson G, Briggs D. High Resolution XPS of Organic Polymers, The Scienta ESCA 300 Database. Vol. 15, Biomaterials. John Wiley & Sons; Chichester [Internet]; 1992. 318 p. Available from: http://www.sciencedirect.com/science?_ob=GatewayURL&_origin=ScienceSearch&_method=citationSearch&_piikey=0142961294900604&_version=1&_returnURL=&md5=65e86f3c03f1fa49fc7247422536913c

[50] Bhattacharyya D, Gleason KK. Single-step oxidative chemical vapor deposition of –COOH functional conducting copolymer and immobilization of biomolecule for sensor application. Chemistry of Materials [Internet]. 2011;**23**(10):2600-2605. Available from: http://pubs.acs.org/doi/abs/10.1021/cm2002397

[51] Park M, Zhang X, Chung M, Less GB, Sastry AM. A review of conduction phenomena in Li-ion batteries. Journal of Power Sources [Internet]. 2010;**195**(24):7904-7929. Available from: http://dx.doi.org/10.1016/j.jpowsour.2010.06.060

[52] Greczynski G, Kugler T, Keil M, Osikowicz W, Fahlman M, Salaneck WR. Photoelectron spectroscopy of thin films of PEDOT – PSS conjugated polymer blend: A mini-review and some new results. Journal of Electron Spectroscopy and Related Phenomena. 2001;**121**(1-3):1-17

[53] Hohnholz D, MacDiarmid AG, Sarno DM, Jones WE Jr. Uniform thin films of poly-3,4-ethylenedioxythiophene (PEDOT) prepared by in-situ deposition. Chemical Communications [Internet]. 2001;**0**(23):2444-2445. Available from: http://xlink.rsc.org/?DOI=b107130k

[54] Xia Y, MacDiarmid AG, Epstein AJ. Camphorsulfonic acid fully doped polyaniline emeraldine salt: In situ observation of electronic and conformational changes induced by organic vapors by an ultraviolet/visible/near-infrared spectroscopic method. Macromolecules. 1994;**27**(24):7212-7214

[55] Kwon IW, Son HJ, Kim WY, Lee YS, Lee HC. Thermistor behavior of PEDOT:PSS thin film. Synthetic Metals. 2009;**159**:1174-1177

[56] Padmalekha KG, Admassie S. Electrochromic, magnetotransport and AC transport properties of vapor phase polymerized PEDOT (VPP PEDOT). Synthetic Metals. 2009;**159**(17-18): 1885-1889

[57] Aleshin AN, Williams SR, Heeger AJ. Transport properties of poly(3,4-ethylenedioxythiophene)/poly(styrenesulfonate). Synthetic Metals [Internet]. 1998;**94**(2):173-177. Available from: http://www.sciencedirect.com/science/article/pii/S0379677997041672

[58] Ouyang J, Xu Q, Chu C-W, Yang Y, Li G, Shinar J. On the mechanism of conductivity enhancement in poly(3,4-ethylenedioxythiophene):poly(styrene sulfonate) film through solvent treatment. Polymer (Guildford) [Internet]. 2004;**45**(25):8443-8450. Available from: http://linkinghub.elsevier.com/retrieve/pii/S0032386104009668

[59] Pacios R, Chatten AJ, Kawano K, Durrant JR, Bradley DDC, Nelson J. Effects of photo-oxidation on the performance of poly [2-methoxy-5- (3′,7′-dimethyloctyloxy)-1,4-phenylene vinylene]:[6,6]-phenyl C 61-butyric acid methyl ester solar cells. Advanced Functional Materials. 2006;**16**(16):2117-2126

[60] Kawano K, Pacios R, Poplavskyy D, Nelson J, Bradley DDC, Durrant JR. Degradation of organic solar cells due to air exposure. Solar Energy Materials & Solar Cells. 2006;**90**(20): 3520-3530

Chapter 8

Polymerizable Materials for Diffractive Optical Elements Recording

Roberto Fernández Fernández,
Víctor Navarro Fuster,
Francisco J. Martínez Guardiola, Sergi Gallego Rico,
Andrés Márquez Ruiz, Cristian Neipp López,
Inmaculada Pascual Villalobos and
Augusto Beléndez Vázquez

Additional information is available at the end of the chapter

http://dx.doi.org/10.5772/intechopen.71511

Abstract

The technologies based on holographic and photonic techniques related to the optical storage and optical processing of information are rapidly evolving. One of the key points of this evolution are the new recording materials able to perform under the most specific situations and applications. In this sense, the importance of the photopolymers is growing spectacularly. This is mainly due to their versatility in terms of composition and design together with other interesting properties such as self-processing capabilities. In this chapter, we introduce the diffractive optical elements (DOE) generation in these materials and some of the most important parameters involved in this process. The deep knowledge of the material is essential to model its behavior during and after the recording process and we present different techniques to characterize the recording materials. We also present a 3D theoretical diffusion model able to reproduce and predict the experimental behavior of the recording process of any kind of DOE onto the photopolymers. The theoretical results will be supported by experimental analysis using a hybrid optical-digital setup, which includes a liquid crystal on silicon display. Besides this analysis, we study a method to improve the conservation and characteristics of these materials, an index-matching system.

Keywords: photopolymer, holography, diffractive optical elements, recording materials

1. Introduction

The holographic recording materials are traditionally used to record holograms, and they have been mainly studied and characterized for holographic applications [1–3]. Along with the evolution of photonics, communications and optical processing of information, the relevance, and capabilities of these materials to store any kind of phase or amplitude diffraction pattern, with good results also for low spatial frequencies, are surfacing [4–6]. At this point, the complete characterization of these materials for low spatial frequencies is decisive to design a material with the optimum characteristics for each specific application [7, 8].

The holographic recording materials change their properties when they are exposed to the light, and there are many examples of them, for example, the photographic emulsions [9], photochromic materials [10], dichromated gelatin [11], photorefractive materials [12], or photopolymers [13].

The last ones, photopolymers, are a very promising option, for example, for the development of holographic memories due to its high resolution and fidelity. The use of this recording material has spectacularly been increased because of its versatility at the time of changing their composition or design [14]. Moreover, they present high reliability, repeatability, and flexibility together with their tunable thickness, self-processing capabilities, and low cost. It is undeniable how those features set the photopolymers as one of the best holographic storage media and a good option for applications inside the diffractive optics and optical processing fields.

In these fields, one of the main drawbacks to be faced is the conservation of the element recorded into the photopolymer. In this sense, along this chapter, we will study the effects and improvements added by a coverplating together with an index-matching system. This system not only improves the lifetime of the material but also affects to the molecules diffusion inside the material.

2. Overview of the materials and DOE recording process

2.1. The recording materials

A photopolymer is a formulation based in an organic polymer sensitive to the light of certain wavelength. The basic formula is made of a sensitizer dye, an initiator to generate free radicals and one or more polymerizable monomers. These components are disposed in a matrix composed of a polymer such as poly(vinyl alcohol) (PVA), sodium polyacrylate, or vinyl chloride (PVC).

The basic way to record a DOE into a photopolymer, a phase media, is by refractive index modulation between the polymerized and nonpolymerized areas, which correspond to illuminated and nonilluminated areas, respectively [15].

A great variety of these materials can be developed based on a given monomer, a support matrix, a dye, and the rest of the components of the solution. These components and their concentration will affect the final properties of the photopolymer, as well as its applications.

The main factor is the binder, because it will determine the range of monomer, dye, and initiator to be used in the final compound.

In this chapter, all the results presented are based on three different materials: one of the most studied photopolymers, based on PVA and acrylamide (AA) as main monomers, which has demonstrated its high linearity and fidelity working in low- and high-spatial frequencies [4, 5]. Despite these good characteristics, the main drawback of this material is its high toxicity, mainly given by AA, which is known to be carcinogenic since many years. Also, it is known that the low environmental compatibility in terms of the low biodegradability of the devices is made of this material [16]. In this sense, our research group has developed a biocompatible photopolymer, called Biophotopol [17], which use sodium acrylate (NaAO) as a main monomer and has demonstrated to have a great dynamic range and high sensitivity together with its high biocompatibility and the main properties of the photopolymers such as the self-processing capabilities and low cost.

The third material used, polymer dispersed liquid crystal (PDLC), demonstrates how the versatility of the photopolymers can be improved by the inclusion of new components, like nanoparticles or dispersed liquid crystal (LC) molecules, in their formula [18, 19]. In this case, thanks to the addition of LC molecules, it is possible to fabricate polymers that change their optical properties by means of an external stimulus, in this case an electrical field. A hologram can be recorded into the material and then modulate its DE on real time thanks to the electrical field applied. The formulations of the different materials are shown in **Table 1**.

2.2. The photopolymerization reaction

The photopolymerization reaction (**Figure 1**) starts when the photopolymers are illuminated by light of certain wavelength, and the monomer starts to polymerize, causing a shrinkage on the illuminated areas, increasing the refractive index and decreasing the monomer concentration of these areas. This chemical compound gradient of the monomer, dye and radical generator causes a diffusion of the different components from the areas where the concentration is high, "dark areas" to the areas where the concentration has decreased, the "illuminated areas." This diffusion causes a decreasing of volume in the dark areas, and therefore, the volume of the illuminated areas is increased by incoming molecules, counteracting the shrinkage due to the polymerization process.

Material: PVA/AA-based photopolymer					
TEA (ml): 2.0	PVA (ml) (8%, w/v): 25	AA (g): 0.84	BMA (g): 0.2	YE (0.8%, w/v) (ml): 0.6	
Material: Biophotopol					
PVA (8%, w/v): 15	NaAO (M): 0.34	TEA (M): 0.15	PRF (M): 1.00×10^{-3}		
Material: PDLC (wt%)					
DPHPA: 48.8	BL036: 29.2	YEt: 0.1	NPG: 1.5	NVP: 16.4	OA: 4.4

Table 1. Formulation of different photopolymers used.

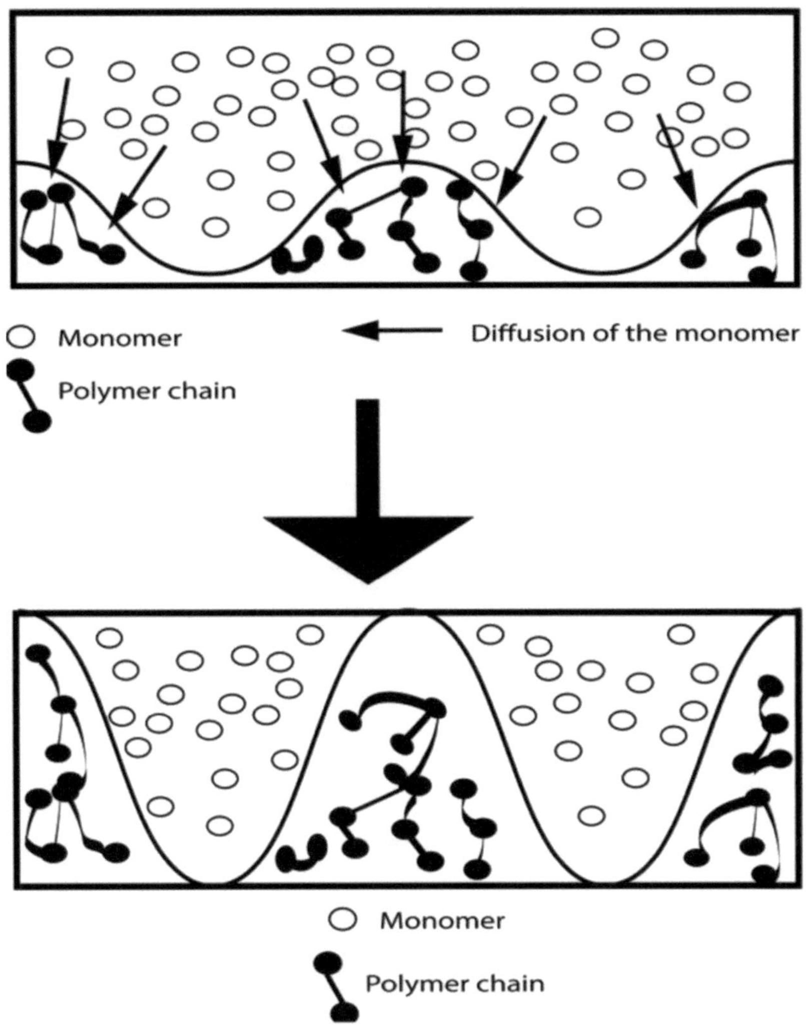

Figure 1. Representation of the formation of the hologram in a photopolymer.

The simplest DOE that can be recorded into a photopolymer is the sinusoidal profile, which can be obtained by the interference of two plane beams, controlling the period of the grating by changing the angle between the two beams. To record more complex DOE, it is necessary to control the shape of the incident beam to generate the desired profile. Nowadays, the advances on the spatial light modulators (SLMs) based on liquid crystal on silicon (LCoS) displays, thanks to the microelectronics technology, are allowing the production of commercial devices with resolutions higher than 4094 × 2464 (4 K) in the same size that of a conventional LCD screen. The use of this device in the optic setups allows the recording of an enormous variety of different complex DOEs with an active control of the function displayed by the LCD screen through a computer.

These complex profiles may suffer a smoothing of the profile due to different factors such as the cutoff frequency of the optical system, the cross talk between pixels, the finite size of the polymer chains, the diffusion of the monomer or the nonlinearity of the recording process. This deviation from the ideal profile depends on the R parameter, which indicates the relative importance of the photopolymerization with respect to the monomer diffusion:

$$R = \frac{DK_g^2}{F_R} \tag{1}$$

where D is the monomer diffusivity, F_R is the polymerization rate, and K_g is the grating number, related with the grating period Λ by means of $K_g = 2\pi/\Lambda$. At the time of modeling the behavior of these materials, there are many other parameters involved in the phase image formation process to consider, starting by the intensity and the fringes visibility, V. The importance of these parameters is analyzed by the relationship between polymerization rate and incident intensity.

$$F_R(t) = k_R(t) \cdot I^{\gamma}(x, z, t) = k_R\left(I_0\left[1 + V\cos\left(K_g x\right)\right]e^{-\alpha(t)z}\right)^{\gamma} \tag{2}$$

The $k_R(t)$ factor is the polymerization parameter, which indicates the speed of the photochemical reaction that takes place in the material and depends on the concentration and type of the substances involved and their media, $I(x,z,t)$ is the recording intensities' distribution, exponentially attenuated with depth due to the dye absorption (Beer's law [20]), α indicates this absorption (it decreases when the dye is consumed), V is the fringes visibility (typically 1), and γ represents the relationship between the polymerization rate and the recording intensity. This parameter usually takes values between 0.5 and 1, corresponding the value 0.5 to more liquid polymerizable systems and 1 to more solid systems.

To model the three-dimensional behavior of the monomer and polymer volume fractions, m and p, respectively, the following general equations can be used, where $\phi^{(m)}$ and $\phi^{(p)}$ are the monomer and polymer concentrations, respectively:

$$\frac{\partial\phi^{(m)}(x,z,t)}{\partial t} = \frac{\partial}{\partial x}D_m(t)\frac{\partial\phi^{(m)}(x,z,t)}{\partial x} + \frac{\partial}{\partial z}D_m(t)\frac{\partial\phi^{(m)}(x,z,t)}{\partial z} - F_R(x,z,t)\phi^{(m)}(x,z,t) \tag{3}$$

$$\frac{\partial\phi^{(p)}(x,z,t)}{\partial t} = F_R(x,z,t)\phi^{(m)}(x,z,t) \tag{4}$$

where D_m is the monomer diffusion inside the material, which decreases with time. To solve these differential equations, there are different methods. In the recording model used through this chapter, we have used the finite-difference method (FDM) to obtain a numerical solution.

Once we have measured the monomer and polymer concentrations, the refractive index of the photopolymeric layer can be calculated as a function of the volume fraction variations of each component using the Lorentz-Lorenz equation [21]:

$$\frac{n^2-1}{n^2+2} = \frac{n_m^2-1}{n_m^2+2}\phi^{(m)} + \frac{n_p^2-1}{n_p^2+2}\phi^{(p)} + \frac{n_b^2-1}{n_b^2+2}\left(1-\phi^{(m_0)}\right) \tag{5}$$

where $\phi^{(m_0)}$ is the average initial value for the monomer volume fraction, n_p is the polymer refractive index, n_m is the monomer refractive index, and n_b is the support matrix refractive index. These two last parameters can be measured using a refractometer, and the value of n_p can be obtained through zero-spatial frequency measurement technique [22].

The linearity in the material response can be studied in very low spatial frequencies to avoid the influence of the monomer diffusion in the diffractive image formation; the R parameter, described in Eq. (1), takes values higher than the ones obtained in holographic regime due to the high value of the spatial period.

The diffraction efficiency (DE) of the different diffracted orders in Fraunhofer domain is given by the Bessel functions [23]. **Figure 2** shows the behavior of the main four diffracted orders as a function of the phase shift. The comparison of this figure and the DE results obtained for certain material gives an approach of the linearity in the response of the material.

2.3. Influence of the index matching

As we mentioned in Section 1, one of the main drawbacks of the photopolymers is their conservation. In this sense, the inclusion of an index-matching system and coverplating not only improves the lifetime and conservation of the materials, but also allow us to measure the internal monomer diffusion [24]. This index-matching system allows us to clarify how the fast changes are measured in uncovered layers. Therefore, some authors observed in PVA/AA materials some fast changes on the surface [4]. They fitted this monomer diffusion around 10^{-7} cm^2/s. Some years late, Close et al. [25], using a coverplating and index-matching agent to

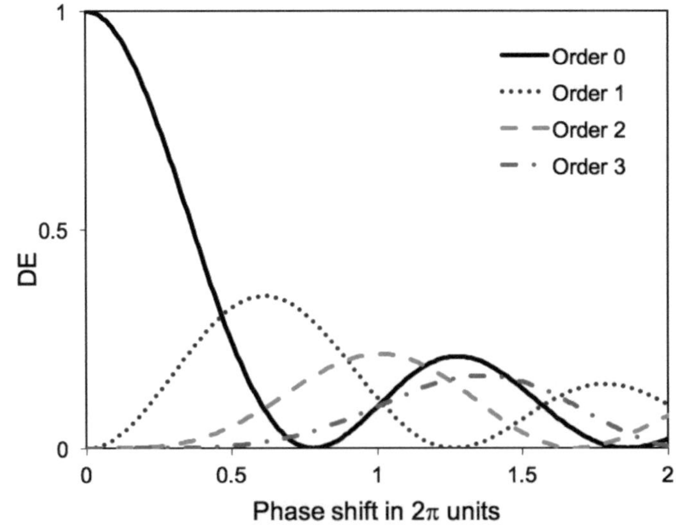

Figure 2. Diffraction efficiency of the main four orders of a sinusoidal grating as a function of the phase depth.

http://dx.doi.org/10.5772/intechopen.71511

avoid the surface changes, measured the monomer diffusion inside the material around 10^{-10} cm^2/s. They also obtained the diffusion of other substances such as acetone inside the material, which lead us to make a distinction between the "apparent" diffusion of the surface, the first one, and the "real" diffusion, the internal monomer diffusion, the second one.

The thickness variations of the material play an important role in the DOEs formation at very low spatial frequency recording [5]. In DOEs recorded in materials without index matching, the transmitted beam has the information of the thickness and refractive index modulation mixed. By using the index-matching system (**Figure 3**), it is possible to study separately the changes produced by the refractive index variation. To achieve the index matching, we must choose a liquid with a refractive index very close to the mean of the polymer refractive index

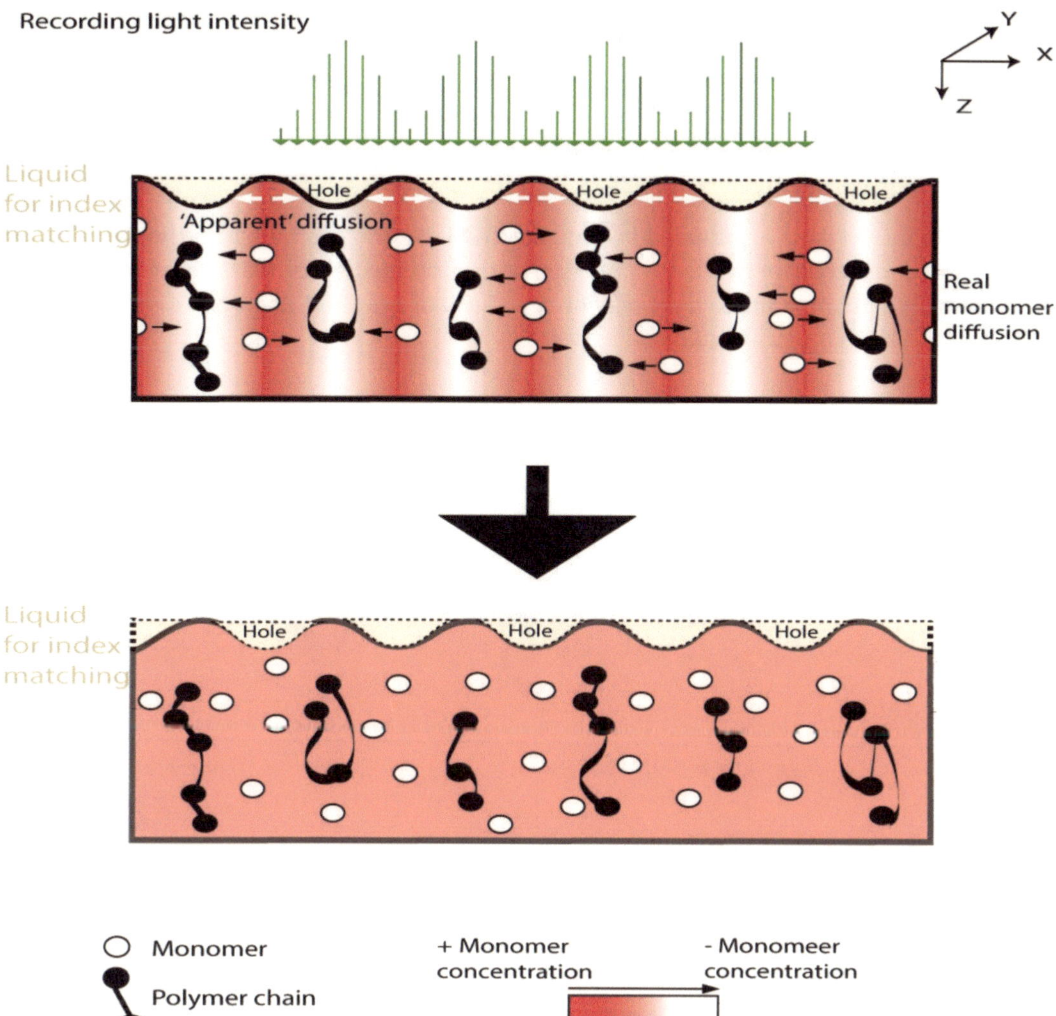

Figure 3. Diagram of a DOE recording using index-matching system. The so-called "apparent" diffusion is due to the recovering surface changes and the "real" diffusion is due to the internal monomer motion.

and not soluble with the photopolymer. When the shrinkage due to the polymerization takes place, this liquid will fill up the holes generated in the surface of the material, minimizing the diffractive effects due to the relief structure.

The value measured for uncovered layers due to this "apparent" diffusion is around 10^{-7} cm^2/s, which is not a suitable value to record complex DOEs with sharp profiles, due to the fast mass transfer, which produces a smoothening of these sharp profiles. For index-matched samples; this value is around 10^{-11} cm^2/s, suitable for complex structures recording. Thanks to the reduction of the mass transfer produced by the sealant, it is possible to record sharp profiles with insignificant smoothening of the profile.

2.3.1. Characterization of the different materials

To analyze the effects of this index-matching system, we use the setup shown in **Figure 4**; in this setup, we can distinguish two beams, the recording beam, provided by a solid-state Verdi laser (Nd-YVO4) with a wavelength of 532 nm (green light), at which the material exhibits maximum absorption, and the analyzing beam, provided by a He-Ne laser at a wavelength of 644 nm, at which the material is not sensitive.

The DOE to be recorded is provided by an LCoS SLM, placed along the recording arm of our setup and sandwiched between two polarizers (P), oriented to produce amplitude-mostly modulation. To obtain a linear response for each level of gray and good contrast, this device

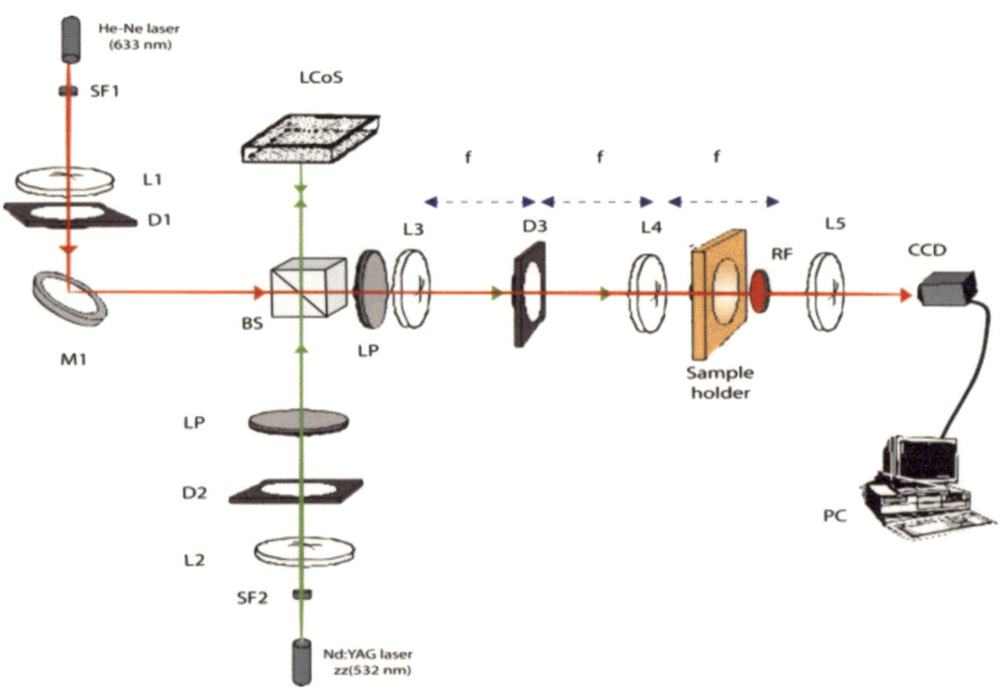

Figure 4. Experimental setup used to record and analyze in real time the DOEs formation on the photopolymers. Di, diaphragm; Li, lens; BS, beam splitter; SFi, spatial filter; LP, linear polarizer; and RF, red filter.

was characterized using the model proposed in [26]. Then, a 4F system images the intensity distribution generated by the SLM onto the recording material. The intensity of the recording beam is 0.25 mW/cm^2.

The analyzing arm, which allows us to study on real time the grating formation onto the photopolymer, was designed to collimate the light incident on the recording material, and a diaphragm (D1) was used to limit the aperture of the collimated beam. A nonpolarizing beam splitter (BS) split the beam in two within the same path. A red filter (RF) was placed behind the recording material to ensure that only the analyzing beam is incident on the CCD camera placed at the end of the setup. To separate the different diffraction orders, we placed a lens behind the material, obtaining the Fraunhofer diffraction pattern on the camera.

In **Figure 5**, the experimental DE of the four main orders for a sinusoidal grating with spatial period of 168 μm is shown. The sample has a thickness of 85 μm. We made two experiments, one without the coverplating and index-matching system (**Figure 5a**) and a second one using the index-matching system and coverplating (**Figure 5b**). It is noticeable how for longer exposure times, greater than 300 s, the phase modulation of both cases looks similar. In samples without index-matching, the recorded phase grating can be understood as the superposition of two phase gratings: a refractive index grating and a relief grating. It can be assumed that the effects of the last one are weaker for long exposure times. At these times, the fast diffusion through the surface in the uncovered sample has mitigated the effects of the relief grating. On the other hand, this relief grating causes that, in the case of the nonindex-matched sample, the maximum DE is achieved after 180 s, three times later that in the index-matched case. This is produced by the decrease in the phase modulation caused by the shrinkage, which is canceled at longer exposure times, as has already been said. In the experiments carried out for gratings with different spatial periods, we have observed that the behavior is similar. This is important for the recording of complex elements with different spatial frequencies mixed in their shapes such as diffractive lenses.

For the holographic regime, the diffusion times are very short and they do not exceed 0.1 s, which greatly hinders its measurement. Thus, to measure the species diffusion in frequencies around 1000 lines/mm, it is necessary to use indirect methods. At very low spatial frequencies,

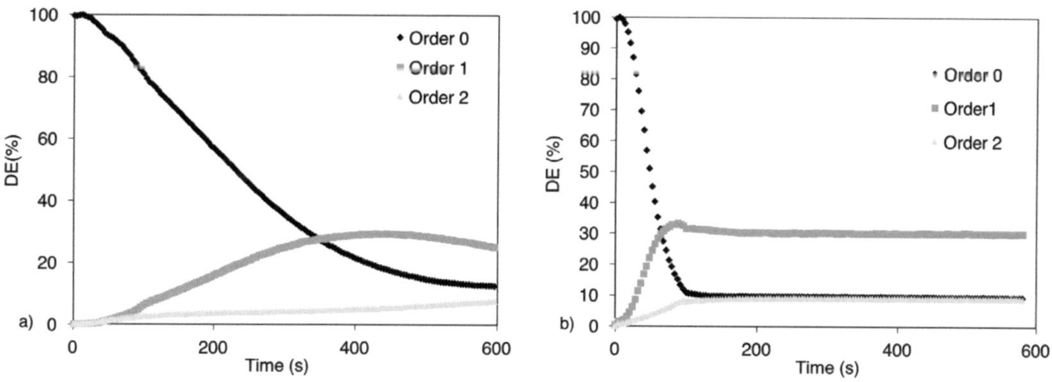

Figure 5. Experimental DE during recording of a grating with spatial period of 168 μm and thickness of 85 μm without index-matching system (a) and with index-matching system (b).

we can measure many diffracted orders to obtain information about the phase profile and its evolution after recording due to the diffusion. An interesting experiment is to measure the evolution of the DOE stored after recording. These variations are called in-dark evolution. In **Figure 6**, it can be appreciated the in-dark evolution of a nonindex-matched sample and an index-matched sample, exposing the sample during 50 s and shutting off the recording laser, analyzing the post exposure evolution of the DOE. For the nonindex-matched samples, the variation of the DEs is very fast in comparison with the index-matched ones. In the last case, it is noticeable how the two first orders remain practically constant after recording. The fitted values of the monomer diffusion in each case can be observed in **Figure 7**. We noted that the phase depth, $\Delta\phi$, becomes time independent after a time range, which goes from several minutes to a few days after exposure. Taking this into account, it is possible to follow the

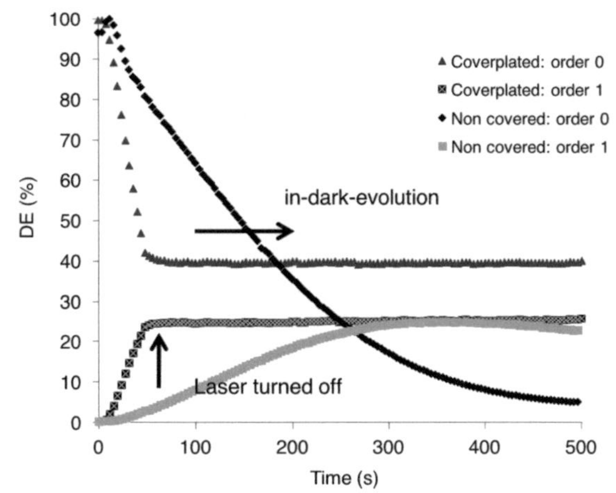

Figure 6. Experimental DE during 50 s recording of a sinusoidal grating of 168 μm in an 85 μm thickness index-matched and nonindex-matched samples and the in-dark evolution.

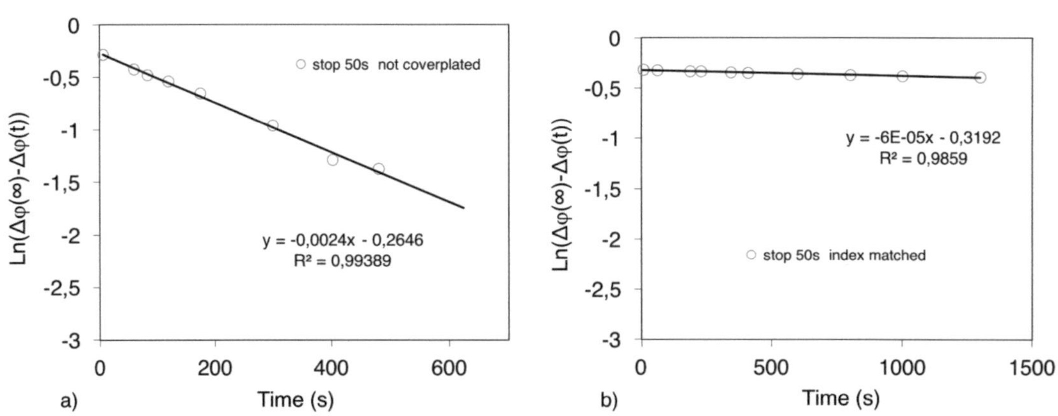

Figure 7. Fitting of the experimental values after 50 s of recording time a nonindex-matched sample (a) and index-matched sample (b).

procedure of [27] to obtain the "apparent" diffusion of this kind of materials, which depends on two variables: Λ, the grating period and τ, the characteristic time of monomer variation.

$$D = \frac{\Lambda^2}{4\pi^2\tau} \tag{6}$$

τ can be calculated by fitting the phase depth of the grating variation with time [6]. To obtain this value, we can use Fick's Law equation to describe species concentration after exposure and the fitting of the temporal variation for Δϕ:

$$\phi^{(i)}(x,t) = \phi_f^{(i)} + \Delta\phi^{(i)}(x)\exp\left(\frac{-t}{\tau}\right) \tag{7}$$

$$\Delta\varphi(\mathrm{x},\mathrm{t}) = V_m\Delta\phi^{(avg)}(x)\left(\exp\left(\frac{-t}{\tau}\right) - 1\right) \tag{8}$$

where the species modulation generated inside the photopolymer is represented by $\Delta\phi^{(i)}(\mathrm{x})$, $\phi_f^{(i)}$ is the average value of the residual species concentration, and $\Delta\phi^{(avg)}$ is the average for all molecules. This corresponds to the point at which monomer diffusion eventually stops due to a uniform monomer distribution. The molar volume of the monomer is represented by V_m and tends asymptotically (in practice in some minutes) to the value:

$$\Delta\varphi(\mathrm{x},\mathrm{t}\to\infty) = -V_m\Delta\phi^{(avg)}(x) \tag{9}$$

Thus, Eq. (8) can be simplified as:

$$\Delta\varphi(\mathrm{x},\mathrm{t}\to\infty) - \Delta\varphi(\mathrm{x},\mathrm{t}) = \Delta\varphi(\mathrm{x},\mathrm{t}\to\infty)\exp\left(\frac{-t}{\tau}\right) \tag{10}$$

And applying logarithm to both sides of the equation, we obtain:

$$\ln\left(\Delta\varphi(x,t\to\infty) - \Delta\varphi(x,t)\right) = \ln\left(\Delta\varphi(x,t\to\infty)\right) - \left(\frac{t}{\tau}\right) \tag{11}$$

The only assumption is that $\Delta\phi$ is proportional to $\Delta^{(avg)}$, which is reasonable for small polymer concentrations.

In **Figure 6**, it can be seen how the slope of the linear fitting for the nonindex-matched sample is almost two magnitude orders higher than the index-matched one. The diffusion measured for the last one is 3.5×10^{-10} cm^2/s, within the expected values for these materials. On the other hand, the "apparent" diffusion measured for the nonindex-matched sample is 1.6×10^{-8} cm^2/s. The differences between using or not the index-matching system are clear together with this separation between "apparent" and "real" diffusion. Using this index-matching system, it is possible to get a phase depth of 2π for PVA/AA samples of 105 μm without using any crosslinker. To get this, 2π phase depth modulation is very important as is needed by many applications related to the complex DOEs recording, such as blazed gratings or diffractive lenses [29, 30].

One of the methods to increase the phase depth is the inclusion of a crosslinker monomer. The role of this component is to increase the chain length and compaction by linking the polyacrylamide chains between them. Therefore, the polymerization rate, shrinkage, and refractive index are increased [30]. With the use of a crosslinker, it is possible to get the 2π phase depth in samples around 85 μm. In **Figure 8**, the effects of using N,N′-methylene-bis-acrylamide (BMA) as a crosslinker agent in the PVA/AA photopolymer formulation is shown. Comparing this figure with **Figure 2**, it is noticeable that the phase depth of $1.4{\cdot}2\pi$ is reached after 200 s of exposure time in a sample of 95 μm thickness, a value substantially lower than the 105 μm needed for materials formulated without this crosslinker. Therefore, including agents such as BMA can drastically improve the material properties at very low spatial frequency recording.

The same analysis had been done for the Biophotopol green photopolymer, studying the effects of the index-matching system. Due to the similarities between this material and the PVA/AA-based one, Biophotopol also presents important relief structures formed during recording, avoiding the separate study of the refractive index distribution generated by polymerization. We have also studied the use of different crosslinker agents such as the already-mentioned BMA or the N,N′-(1,2-dihydroxyethylene) bisacrylamide (DHEBA), a highly environmental compatible crosslinker, which is also able to dissolve the NaAO, increasing the utility life of the samples up to 10 times, preventing the monomer crystallization [31]. **Figure 9** shows the experimental diffraction efficiency of a Biophotopol + DHEBA sample. In general, the results obtained are very similar to the ones obtained for the other photopolymers. In this case, the maximum phase depth obtained was around $1.4\,\pi$ for a 90 μm sample. Therefore, assuming a linear behavior with thickness, a sample of around 128 μm will be required to achieve a phase depth of 2π. Regarding the monomer diffusion, the values obtained were around 10^{-10} cm^2/s, a bit slower compared to PVA/AA material.

At this point, it is also interesting to show the capabilities of the PDLC material to work with low spatial frequencies and develop tunable DOEs by means of an electrical field. The use of this material is very common for holographic regime, where it presents a high value of refractive index modulation providing DEs very close to 100% for optical thicknesses around 10 μm. The main drawback of this material is its high scattering. Regarding this factor, it is

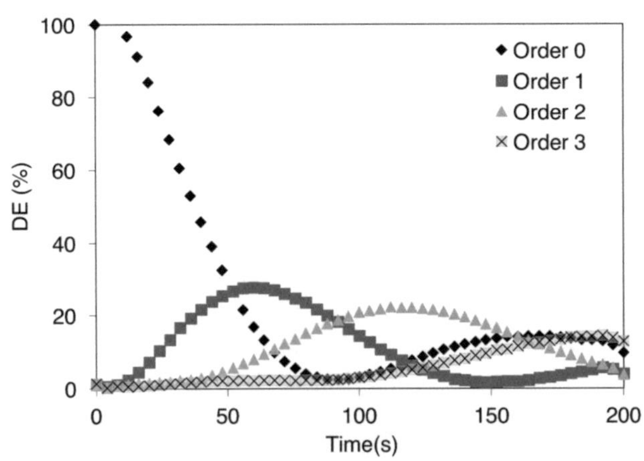

Figure 8. Experimental DE for the recording of a sinusoidal grating of 168 μm during 200 s of exposure time in an AA/PVA material with 95 μm thickness without the index-matching and coverplating system.

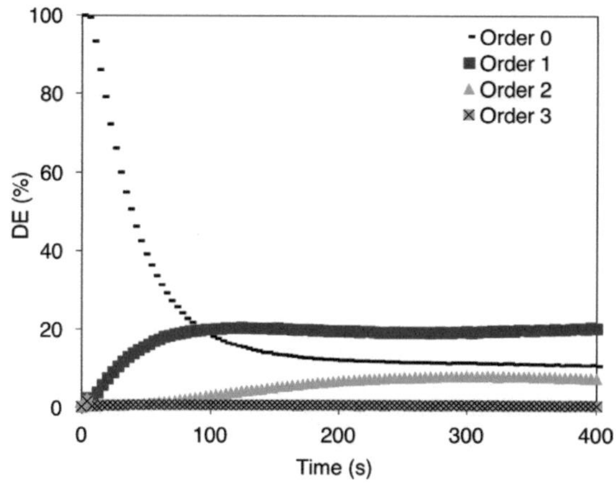

Figure 9. Experimental DE for the recording of a sinusoidal grating of 168 μm during 400 s of exposure time in an Biophotopol + DHEBA material with 90 μm thickness.

important to remark that this family of materials is always enclosed between indium tin oxide (ITO) glasses that make impossible to use the index-matching technique used in the other two materials. Thus, in this case, it is impossible to distinguish between "real" and "apparent" diffusion. The results obtained for the recording of a 168 μm period sinusoidal grating into a 30 μm thickness layer are shown in **Figure 10**. It can be seen how the phase depth of 2π can be achieved, but despite this, the fast values of diffusion, around 10^{-8} cm^2/s, together with the impossibility to distinguish the internal monomer diffusion from the "apparent" one make its selection difficult as a candidate to record low-frequency DOEs.

2.4. Simulation of the behavior of the materials

A model able to simulate the behavior of the material is a powerful tool to obtain the desired material or illumination parameters to fabricate a particular DOE. We have already introduced our recording model in Section 2.2. Based on the solving of the differential equations that

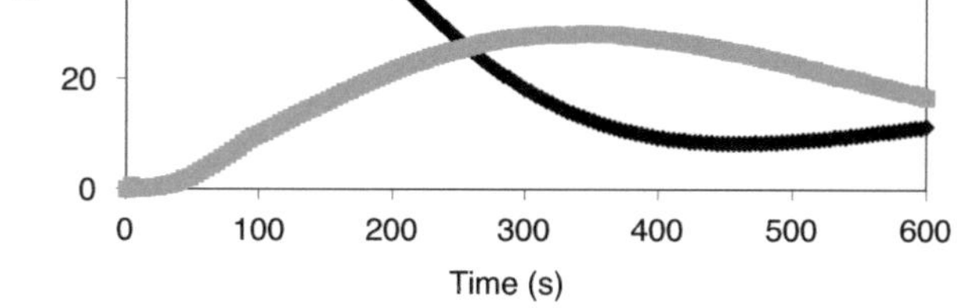

Figure 10. Experimental DE for the recording of a sinusoidal grating of 168 μm during 600 s of exposure time in a PDLC material with 30 μm thickness.

represents the behavior of the monomer and polymer volume fractions, it is possible to predict with fidelity how the different materials will behave during the DOEs recording. This model also includes the effects of the holes produced in the surface during polymerization process and analyzes both the hole and monomer diffusion separately [32].

Holes generation and its diffusion role in the model can be taken into account adding the following equations to the previously introduced, in Section 2.2:

$$\frac{\partial\phi^{(h)}(x,t)}{\partial t} = \frac{\partial}{\partial x} D_h(t)\frac{\partial\phi^{(h)}(x,z,t)}{\partial x} - K_h(x,t)\phi^{(m)}(x,t) \tag{12}$$

where $\phi^{(h)}$ represents the holes volume fraction and K_h is the holes rate generation, proportional to F_R, and D_h is the diffusion constant for holes. We assume that the holes are concentrated close to the surface, and their motion is along x direction, with the grating vector parallel to x-axis.

To check the capability of the recording model to simulate both the recording process on the material and the role of the sealant in this process, we recorded sinusoidal gratings in a PVA/AA material. The results obtained were compared with the simulation results, introducing in the model the parameters of the analyzed material. In **Figure 11** both the experimental results are shown that are compared with the simulation results of the DE of the main four orders of a sinusoidal grating with a period of 168 μm recorded into a PVA/AA material of 80 μm without (a) and with index matching (b). In both cases, the good agreement can be appreciated between the model and the experimental results. The results obtained are in line with the results shown previously, and the index-matched sample takes less time to reach the maximum DE due to the mitigation of the effects produced by the thickness variations.

The model is also capable to faithfully reproduce the postexposure evolution of the material in both index-matched and nonindex-matched samples. In **Figure 12**, it can be seen, apart from the good agreement between experimental and simulation results, how for the nonindex-

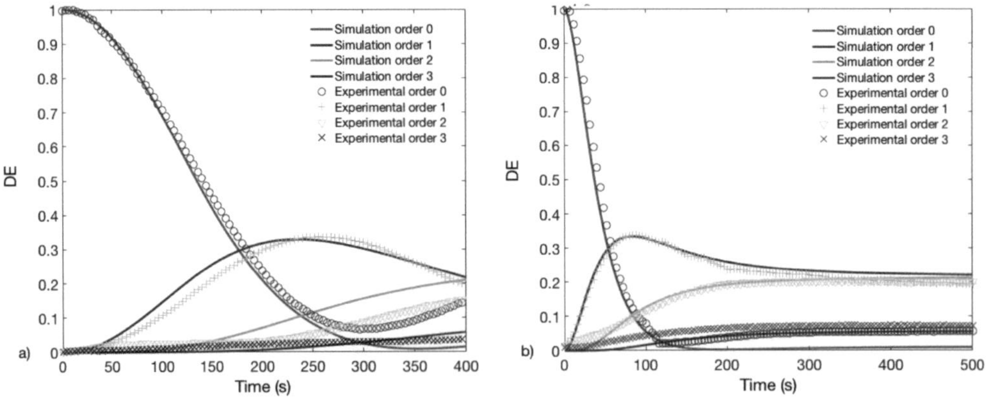

Figure 11. Comparison between experimental and simulated results of the main four orders of DE as a function of time for the recording of a sinusoidal grating of 168 μm in a PVA/AA material with 80 μm thickness, without index matching (a) and with it (b).

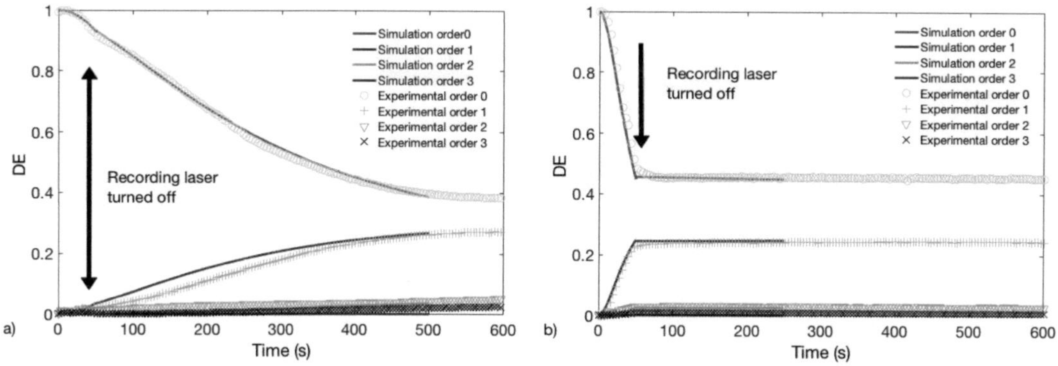

Figure 12. Comparison between experimental and simulated results of the postexposure evolution of the main four orders of DE as a function of time for the recording of a sinusoidal grating of 168 μm in a PVA/AA material with 80 μm thickness, without index matching (a) and with it (b).

matched sample (a) the DE continues changing after shutting of the recording beam due to the hole diffusion through the surface. These effects are not present in the index-matched sample (b), in which the DE evolution stops with the shutdown of the laser.

3. Recording of complex DOEs

3.1. Recording of blazed gratings

The lower monomer diffusion shown by the materials with index-matching system together with their capabilities of reaching 2π phase depth makes us able to explore the recording of more complex DOEs. One of these complex profiles is the blazed grating, a sharp profile with abrupt changes, which has many applications in communications and theoretically can reach DEs of 100% [28].

Taking into account the previous results of monomer diffusion, we attempt to store this kind of DOE in PVA/AA and Biophotopol materials. The value of D_0 tested for PDLC materials avoids the generation of sharp profiles on them. The recording model used is the one described in Section 2.2. In this case, the light intensity distribution during the recording process can be written as follows:

$$I = \frac{I_0}{\frac{1}{f_s \cdot \Delta_x} + 1} \tag{13}$$

where f_s is the period of the grating and I_0 is the maximum recording intensity.

The experimental setup has been described previously in Section 2.3.1. We placed the CCD camera in the material plane, and the image at this place and the intensity distribution of this image are shown in **Figure 13**. This figure shows a smoothening of the abrupt edges of the profile due to the low-pass filtering that the experimental setup introduces, especially due to

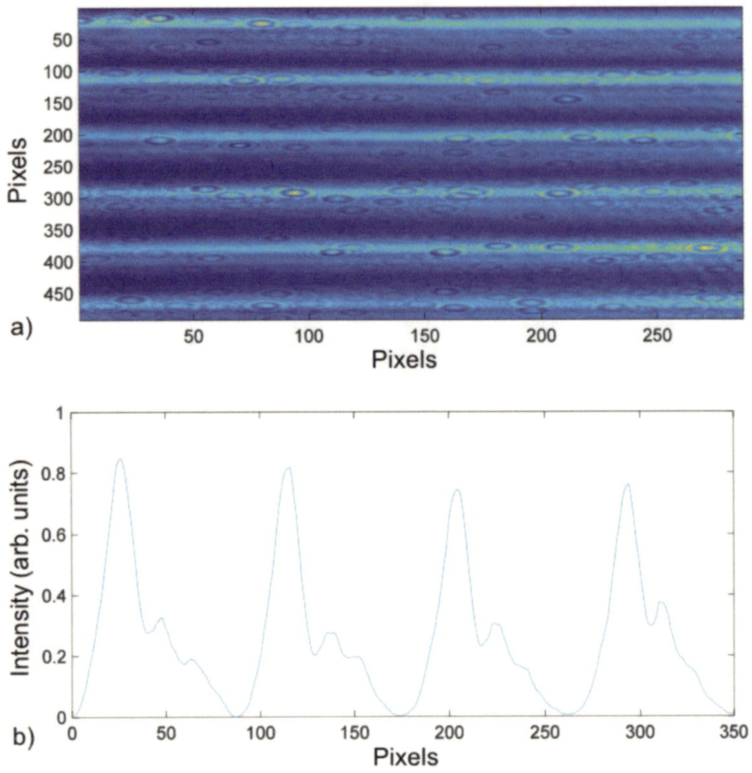

Figure 13. Image of a 672-μm blazed grating provided by the LCoS captured by the CCD at the material plane (a) and intensity profile across a vertical line of this image (b).

the diaphragm (D3) placed to eliminate the pixilation of the LCoS screen of the SLM. To improve the accuracy of the recording model, we introduced this recording intensity on it so that the model takes into account the low-pass filtering introduced by the experimental setup.

To analyze the response of the different materials and the capability of our model to predict the behavior of these materials, we perform different period gratings recording. First, we simulated the recording of this kind of DOE using our model to have an idea of the index-matching influence in both materials for different periods. Once we had a theoretical idea of the influence of index matching and period, we compared the simulations with the experimental results. **Figure 14** shows the simulated and the experimental DEs of a blazed grating of 672 μm recorded in a PVA/AA photopolymer 90 μm thick and a 336 μm one recorded in a Biophotopol photopolymer of 90 μm. It can be seen how the DE of the first order reaches a maximum of almost 70% after an exposure time of 150 s in the PVA/AA photopolymer and almost 55% in the Biophotopol one. Considering the low-pass filtering introduced by the setup, these are good results because we achieved near the maximum value of DE achievable taking into account this low-pass filtering. It is also noticeable that there is a good agreement between the simulation and the experimental results. The lower value of DE obtained for Biophotopol is due to the lower values of k_r and n_p of this family of photopolymers and can

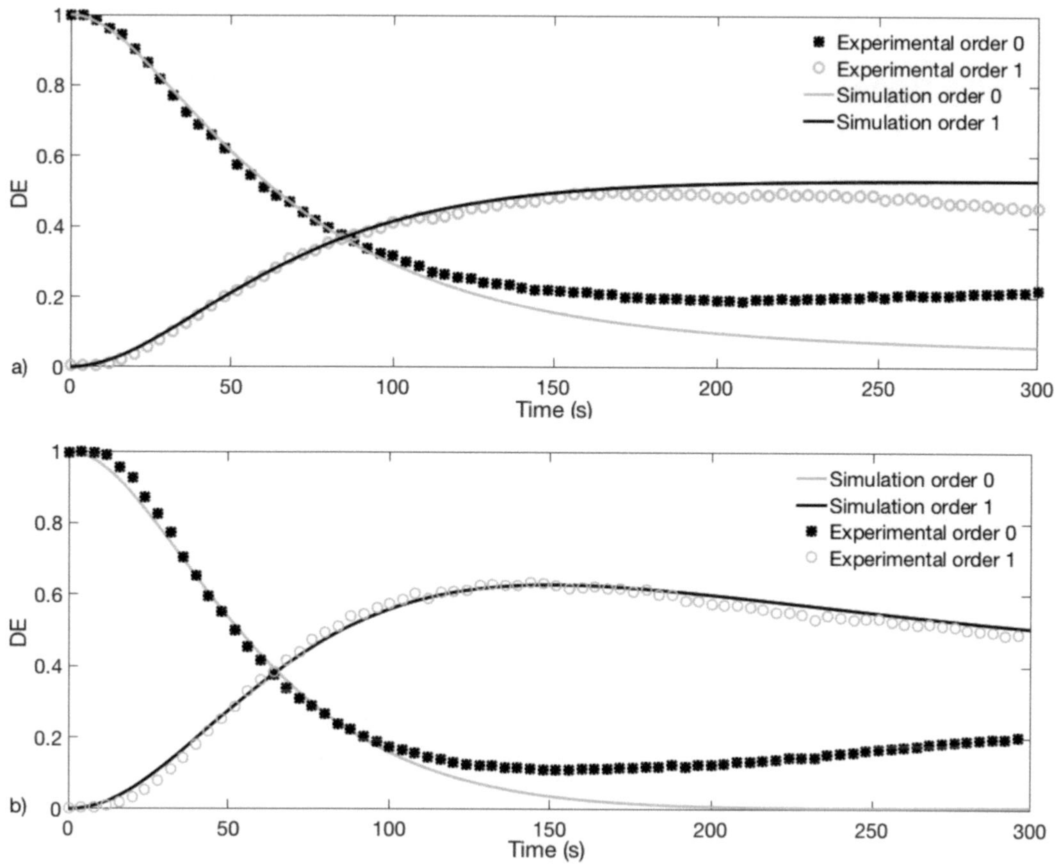

Figure 14. Comparison of the simulated and experimental DE of a 672 μm blazed grating recorded in a PVA/AA material (a) and a 336 μm blazed grating recorded in Biophotopol (b), both during an exposure time of 300 s.

be increased using thicker samples or higher concentrations of crosslinker in the final solution. For shorter period gratings recorded in the same kind of material, it is also appreciable a reduction of the maximum DE reached. This is due to the higher influence of the low-pass filtering at shorter spatial periods together with a higher diffusion.

In both cases and in the rest of the analysis carried out, the results present similarity independent of the spatial frequency of the grating, showing the low influence of the monomer diffusion at these spatial frequencies.

The recording model allows us to probe the capability of the materials to reach the 100% of DE without considering the low-pass filtering introduced by the experimental setup. This simulation is shown in **Figure 15**, where it can be seen how a DE of 100% could be reached for PVA/AA-based material and a DE of 97% for the Biophotopol one. The low-pass filtering introduced by the experimental setup means a reduction of over 20% of DE for both materials. This value is achieved at less exposure time than in the experiments by both materials due to the ideal recording intensity. It is also noticeable how in this ideal simulation, the PVA/AA-based

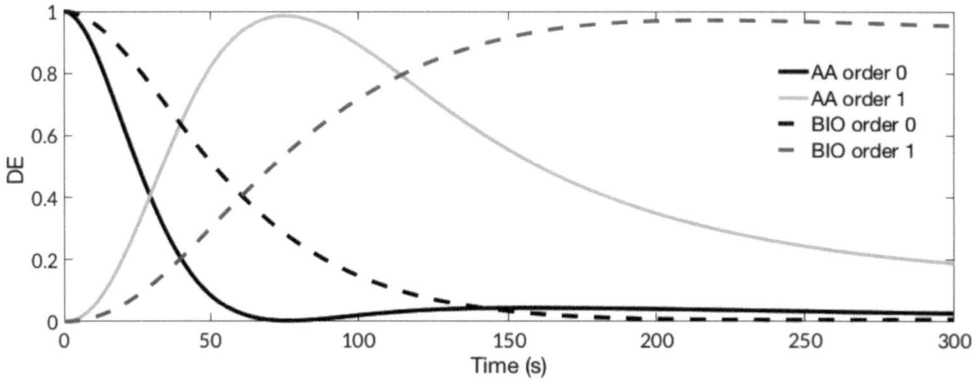

Figure 15. Simulation results of the DE as a function of time for PVA/AA and Biophotopol materials without taking into account the low-pass filtering introduced by the experimental setup.

material takes less time to reach the maximum DE than the Biophotopol due to the differences in n_p and k_r.

3.2. Diffractive lenses

The capabilities of the materials working with a complex profile such as blazed gratings have been proved obtaining near the maximum DE achievable taking into account the low-pass filtering introduced by the setup. Also, the recording model has demonstrated its fidelity reproducing the recording of this element in the different materials, with different spatial periods and considering the index-matching system and low-pass filtering.

Going further, another example of complex DOE to evaluate the model and materials capabilities is diffractive lenses. This DOE includes different spatial periods in its shape, being critical to avoid the smoothening of the profile and to achieve similar behavior for the different spatial periods.

The study of the recording of this kind of DOE has been made only for PVA/AA materials and Biophotopol for the same reasons of the blazed gratings. It is also important to remark that it is necessary to adapt the recording model to make it able to reproduce the cylindrical and spherical lenses formation in these materials. The two-spatial dimension equation shown in Section 2.2 can be applied to reproduce the cylindrical lenses behavior; nevertheless, it is necessary to add a new dimension, the "y" variable dimension, to simulate the case of a spherical lens recording [30]. Eqs. (3) and (4) remain as follows:

$$\frac{\partial \phi^{(m)}(x,y,z,t)}{\partial t} = \frac{\partial}{\partial z} D_m(t) \frac{\partial \phi^{(m)}(x,y,z,t)}{\partial z} + \frac{\partial}{\partial y} D_m(t) \frac{\partial \phi^{(m)}(x,y,z,t)}{\partial y} + \frac{\partial}{\partial x} D_m(t) \frac{\partial \phi^{(m)}(x,y,z,t)}{\partial x} - F_R(x,z,t)\phi^{(m)}(x,z,t) \tag{14}$$

$$\frac{\partial \phi^{(p)}(x,y,z,t)}{\partial t} = F_R(x,y,z,t)\phi^{(m)}(x,y,z,t) \tag{15}$$

The recording intensity distribution is generated by the LCoS SLM in the amplitude regime, which, as in the previously shown experiments, is projected on the material to generate the corresponding phase element. This intensity distribution is defined in Eq. (16), where the phase depends on the quadratic value of the distance between the point and the lens centrum. This equation then is wrapped to 2π and normalized to the maximum value of intensity, I_0.

$$/(x,y) = exp\left[j\frac{\pi}{\lambda f}(x^2+y^2)\right] \tag{16}$$

In this equation, f is the focal length and λ represents the light wavelength. The intensity distribution of a 50 cm focal length lens generated using the SLM is shown in **Figure 16**.

The experimental setup used to record diffractive lenses is also the one already shown in **Figure 4**, with the particularity that in this case, L5 is not present as the lens recorded into the photopolymer is responsible for the focusing of the 633 nm wavelength beam. We imaged the point spread function (PSF) generated by this lens onto the CCD camera, controlling the magnification of the setup by means of the 4F system. The camera is also placed in the material plane to evaluate the intensity pattern imaged on the material. The image taken at this plane and the intensity profile across a horizontal line passing through the center of the lens are shown in **Figure 17**. We can see the characteristic ring structure with a decreasing spatial period as we move away from the center of the diffractive lens. Also, in the intensity profile image, it is noticeable the improvement of the profile with respect to ones presented using a transmissive LCD as SLM with a pixel size of 44 μm [33].

Using the recording model, it is possible to predict the refractive index distribution generated by the incident beam modulated by the SLM. Once we had obtained this refractive index distribution, we can apply the Fresnel propagation [34] to calculate the intensity distribution as a function of time. Thus, we can analyze the focalization of the lens and the optimum recording time to obtain a good focalization. In **Figure 18**, the results of the recording simulation can be seen for the recording of a 50 cm diffractive spherical lens in a PVA/AA material

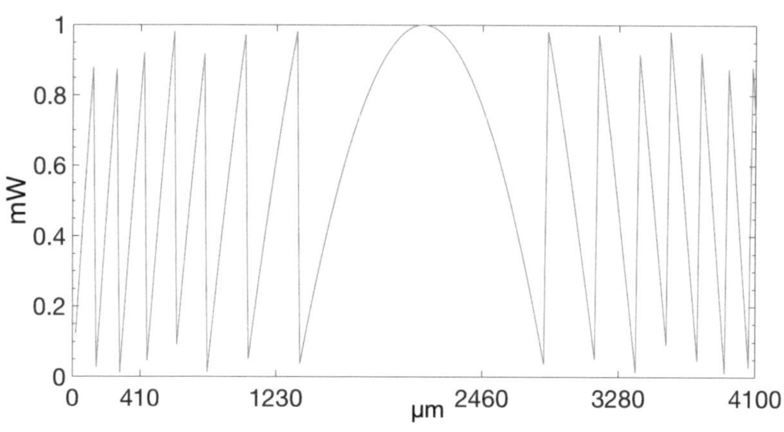

Figure 16. Horizontal cut of the theoretical intensity distribution of a 50 cm focal lens to be projected onto the photopolymer by the SLM.

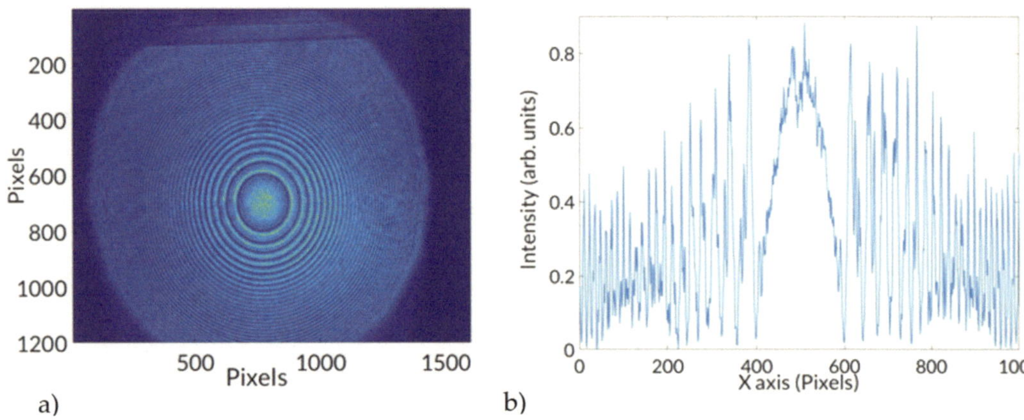

Figure 17. Image provided by the LCoS SLM at the material plane captured by the CCD camera (a) and intensity profile across a horizontal line passing through the center of the lens (b).

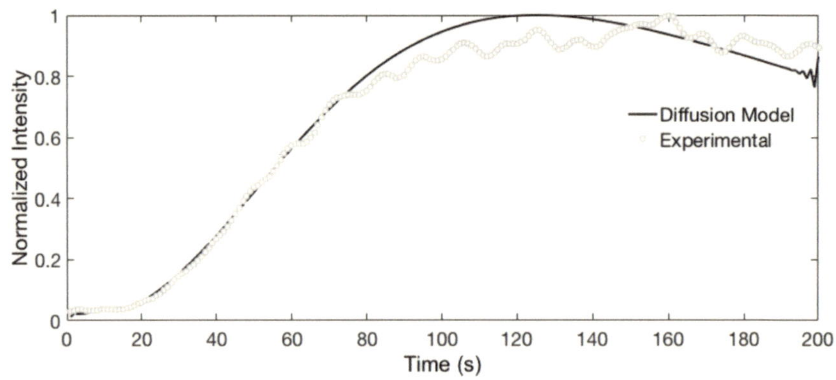

Figure 18. Intensity at the focal point as a function of time for a spherical lens f = 50 cm in a PVA/AA material of 95 μm.

with index matching compared to the experimental results, specifically the intensity at the focal point as a function of time. At this point, it is important to add that the simulated behavior exhibited by cylindrical and spherical lenses in the photopolymers used and different focal lenses is similar. Thus, the results shown correspond to the spherical one. In the figure, the good agreement is noticeable between the simulation and the experimental results, together with the good focalization power of the lenses. The model is also capable to predict the optimum recording time for the lenses in the different materials, as it will be shown in the following figures, and the decrease in the focal intensity. This occurs when the exposure time is longer than the optimum and the phase modulation overcomes 2π, as we chemically design the materials to obtain phase depth saturation slightly higher than 2π for the physical thickness used.

Thanks to the small pixel size of the LCoS screen of the SLM, it is possible to attempt to store shorter focal lenses, for example, 13 cm, to study the influence of the focal on the lenses recording. **Figure 19** shows the comparison between the simulation and the experimental measurements of the intensity at the focal point as a function of time for two different focal lenses recorded in Biophotopol material. In this case, there is also a good agreement between the simulation and the experimental results. Both focal lenses present similar behavior than the one exhibited by the lens recorded in PVA/AA material. The lens with shorter focal takes less time to focalize, maybe due to the higher influence of the monomer diffusion.

3.2.1. *Influence of the material parameters*

At the beginning of the chapter, the importance of different parameters in the recording process was remarked. One of the most important aspects of having a recording model that takes into account many of these parameters is that we can study separately the influence of each one of them in the final DOE recorded. In this section, we will discuss the influence of one of these parameters, the internal monomer diffusion (D_0), as a sample of the model's capability. The influence of other parameters, such as the relationship between intensity and polymerization (γ) and the influence of the depth attenuation (α), can be consulted in Ref. [29]. We have studied the variation of each one of these parameters keeping the rest constant. This gives us the idea of which parameters are more important for the lens formation.

To check the influence of D_0, we have simulated different lenses with a range of monomer diffusivities from 3×10^{-8} to 3×10^{-11} cm^2/s and studied how the diffusion affects in the focal plane. **Figure 20** shows the intensity at the focal point for different internal monomer diffusivities. In this case, the intensity at the focal point for the two smallest values does not show any important difference. On the other hand, for both highest values of diffusivity, we appreciate a clear variation as it was expected. The increase in the monomer diffusivity has influence for these spatial frequencies for values higher than 3×10^{-10} cm^2/s. Below this value, the influence

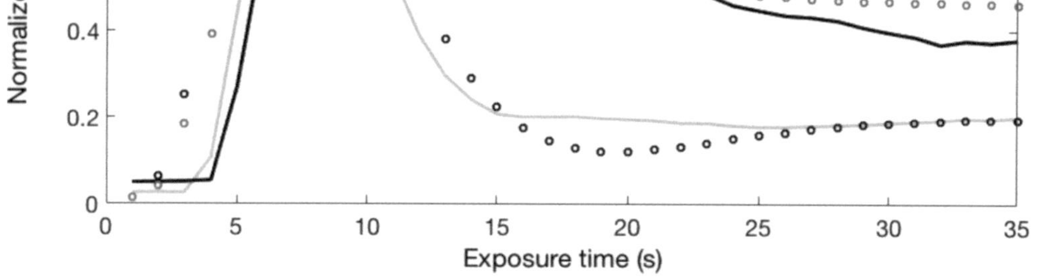

Figure 19. Comparison of the experimental results and the simulation of the intensity at the focal point for two different focal lengths (13 and 60 cm) of diffractive lenses recorded in Biophotopol material of 140 μm thickness.

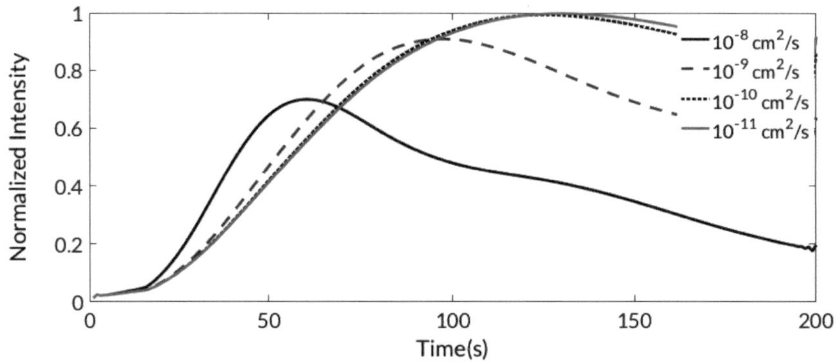

Figure 20. Intensity at the focal point as a function of time for different internal monomer diffusivities.

is low in the lens formation. Therefore, we assumed that for the PDLC material presented in a previous section, the results will not be as good as for the PVA/AA and Biophotopol materials.

4. Conclusions

In this chapter, we have presented a method to record low-frequency diffractive optical elements into photopolymers. This process is influenced by many parameters that we have introduced in a three-dimensional diffusion model to predict the phase image formation. Using this model together with different experimental measurements, we have developed an analysis of the requirements needed to achieve a phase modulation range of 2π in different families of materials. The results show the effectiveness and versatility of the recording model used. Thanks to this model, it is possible to predict the experimental behavior of the recording of any kind of DOE in the different photopolymers. The effectiveness of the model was validated by the experimental work carried out and based in the inclusion of an LCoS SLM. This device allows us to store any kind of DOE selected dynamically and analyze the influence of the different material properties during the recording.

Together with that, we have shown the effects of using the index-matching system, which apart from improving the conservation and lifetime of the recorded DOE, let us differentiate between the diffusion in the surface of the photopolymer and the internal diffusion. The lower values of diffusion obtained in the index-matched materials can be exploited to record sharp DOEs without significant smoothening of the refractive index profile.

Acknowledgements

We want to acknowledge the financial support from the Spanish Ministerio de Trabajo y Competitividad under projects FIS2014-56100-C2-1-P and FIS2015-66570-P and by the Generalitat Valenciana of Spain under project PROMETEOII/2015/015.

Author details

Roberto Fernández Fernández*, Víctor Navarro Fuster, Francisco J. Martínez Guardiola, Sergi Gallego Rico, Andrés Márquez Ruiz, Cristian Neipp López, Inmaculada Pascual Villalobos and Augusto Beléndez Vázquez

*Address all correspondence to: roberto.fernandez@ua.es

University Institute of Physics Applied to Sciences and Technologies, University of Alicante, Alicante, Spain

References

[1] Weiser MS, Bruder FK, Fäcke T, Hönel D, Jurbergs D, Rölle T. Self-processing, diffusion-based photopolymers for holographic applications. Macromolecular Symposia. 2010;**296**(1):133-137. DOI: 10.1002/masy.201051020

[2] Lechner MD. Photopolymers for optical memories and waveguides. In: Kuzmany H, Mehring M, Roth S, editors. Electronic Properties of Polymers and Related Compounds. Berlin, Heidelberg: Springer; 1985. DOI: 10.1007/978-3-642-82569-9_55

[3] Coufal HJ, Psaltis D, Sincerbox DT, editors. Holographic Data Storage. Berlin, Heidelberg: Springer; 2000. DOI: 10.1007/978-3-540-47864-5

[4] Baveba T, Naydenova I, Martin S, Toal V. Method for characterization of diffusion properties of photopolymerisable systems. Optics Express. 2008;**16**(12):8487-8497. DOI: 10.1364/OE.16.008487

[5] Close CE, Gleeson MR, Sheridan JT. Monomer diffusion rates in photopolymer material. Part I. Low spatial frequency holographic gratings. Journal of the Optical Society of America B. 2011;**28**(4):658-666. DOI: 10.1364/JOSAB.28.000658

[6] Gallego S, Márquez A, Ortuño M, Francés J, Marini S, Pascual I, Beléndez A. Surface relief model for without coverplating photopolymers. Optics Express. 2011;**19**(11):10896-10906. DOI: 10.1364/OE.19.010896

[7] Gallego S, Márquez A, Martínez-Guardiola FJ, Riquelme M, Fernández R, Pascual I, Beléndez A. Linearity in the response of photopolymers as optical recording media. Optics Express. 2013;**21**(9):10995-11008. DOI: 10.1364/OE.21.010995

[8] Zhao G, Mouroulis P. Diffusion model of hologram formation in dry photopolymers materials. Journal of Modern Optics. 1994;**41**(10):1929-1939. DOI: 10.1080/09500349414551831

[9] Bjelkhagen HI. Silver-Halide Recording Materials. 2nd ed. Berlin, Heidelberg: Springer; 1995. 441 p. DOI: 10.1007/978-3-540-70756-1

[10] Duncan RC, Staebler DL. Inorganic photochromic materials. In: Smith HM, editor. Holographic Recording Materials. Topics in Applied Physics. Vol. 20. Berlin, Heidelberg: Springer; 1977. p. 133-160. DOI: 10.1007/3-540-08293-X_45

[11] Chang BJ, Leonard CD. Dichromated gelatin for the fabrication of holographic optical elements. Applied Optics. 1979;**18**(14):2407-2417. DOI: 10.1364/AO.18.002407

[12] Bradley DDC, Gelsen OM. Comment on observation of the photorefractive effect in a polymer. Physical Review Letters. 1991;**67**:2589. DOI: 10.1103/PhysRevLett.67.2589

[13] Günter P. Coherent light amplification and optical phase conjugation with photorefractive materials. Journal de Physique Colloques. 1983;**44**:C2–141-C2–147. DOI: 10.1051/jphyscol:1983219

[14] Calvo ML, Cheben P. Photopolymerizable sol–gel nanocomposites for holographic recording. Journal of Optics A: Pure and Applied Optics. 2009;**11**(2):024009. DOI: 10.1088/1464-4258/11/2/024009

[15] Kveton M, Fiala P, Havranek A. Polymer holography in acrylamide-based recording material. In: Rosen J, editor. Holography, Research and Technologies. InTech; 2011. DOI: 10.5772/14564

[16] Hashimoto K, Aldridge WN. Biochemical studies on acrylamide, a neurotoxic agent. Biochemical Pharmacology. 1970;**19**(9):2591-2604. DOI: 10.1016/0006-2952(70)90009-2

[17] Ortuño M, Fernández E, Gallego S, Beléndez A, Pascual I. New photopolymer holographic recording material with sustainable design. Optics Express. 2007;**15**(19):12425-12435. DOI: 10.1364/OE.15.012425

[18] Hata E, Mitsube K, Momose K, Tomita Y. Holographic nanoparticle-polymer composites based on step-growth thiol-ene photopolymerization. Optical Materials Express. 2011;**1**(2):207-222. DOI: 10.1364/OME.1.000207

[19] Suzuki N, Tomita Y. Silica-nanoparticle-dispersed methacrylate photopolymers with net diffraction efficiency near 100%. Applied Optics. 2004;**43**(10):2125-2129. DOI: 10.1364/AO.43.002125

[20] Gallego S, Ortuño M, Neipp C, Márquez A, Kelly JV, Sheridan JT, Beléndez A, Pascual I. 3D behaviour of photopolymers as holographic recording material. In: Proc. SPIE 6252, Holography 2005: International Conference on Holography, Optical Recording, and Processing of Information; 9 June 2006; Varna, Bulgaria, SPIE; 2006. p. 62520B. DOI: 10.1117/12.676525

[21] Aubrecht I, Miler M, Koudela I. Recording of holographic diffraction gratings in photopolymers: Theoretical modelling and real-time monitoring of grating growth. Journal of Modern Optics. 1998;**45**(7):1465-1477. DOI: 10.1080/09500349808230641

[22] Gallego S, Márquez A, Méndez D, Neipp C, Ortuño M, Álvarez M, Fernández E, Beléndez A. Real-time interferometric characterization of a polyvinyl alcohol based photopolymer at the zero spatial frequency limit. Applied Optics. 2007;**46**(30):7506-7512. DOI: 10.1364/AO.46.007506

[23] Hariharan P. Optical Holography: Principles, Techniques and Applications. 2nd ed. Cambridge: Cambridge University Press; 1996. DOI: 10.1017/CBO9781139174039

[24] Fernández R, Gallego S, Márquez A, Francés J, Martínez FJ, Beléndez A. Influence of index matching on AA/PVA photopolymers for low spatial frequency recording. Applied Optics. 2015;**54**(11):3132. DOI: 10.1364/AO.54.003132

[25] Close CE, Gleeson MR, Mooney DA, Sheridan JT. Monomer diffusion rates in photopolymer material. Part II. High-frequency gratings and bulk diffusion. Journal of the Optical Society of America B. 2011;**28**(4):842-850. DOI: 10.1364/JOSAB.28.000842

[26] Martínez FJ, Márquez A, Gallego S, Ortuño M, Francés J, Pascual I, Beléndez A. Predictive capability of average Stokes polarimetry for simulation of phase multilevel elements onto LCoS devices. Applied Optics. 2015;**54**(6):1379-1386. DOI: 10.1364/AO.54.001379

[27] Gallego S, Márquez A, Marini S, Fernández E, Ortuño M, Pascual I. In dark analysis of PVA/AA materials at very low spatial frequencies: Phase modulation evolution and diffusion estimation. Optics Express. 2009;**17**(20):18279-18291. DOI: 10.1364/OE.17.018279

[28] Fernández R, Gallego S, Márquez A, Francés J, Navarro-Fuster V, Beléndez A. Blazed gratings recorded in absorbent photopolymers. Materials. 2016;**9**(3):195. DOI: 10.3390/ma9030195

[29] Fernández R, Gallego S, Márquez A, Francés J, Navarro-Fuster V, Pascual I. Diffractive lenses recorded in absorbent photopolymers. Optics Express. 2016;**24**(2):1559-1572. DOI: 10.1364/OE.24.001559

[30] Andrzejewska E. Photopolymerization kinetics of multi-functional monomers. Progress in Polymer Science. 2001;**26**(4):605-665. DOI: 10.1016/S0079-6700(01)00004-1

[31] Fernández R, Gallego S, Francés J, Pascual I, Beléndez A. Characterization and comparison of different photopolymers for low spatial frequency recording. Optical Materials. 2015;**44**(1):18-24. DOI: 10.1016/j.optmat.2015.02.025

[32] Gallego S, Fernández R, Márquez A, Ortuño M, Neipp C, Gleeson MR, Sheridan JT, Beléndez A. Two diffusion photopolymer for sharp diffractive optical elements recording. Optics Letters. 2015;**40**(14):3221-3224. DOI: 10.1364/OL.40.003221

[33] Marquez A, Gallego S, Ortuño M, Fernandez E, Alvarez ML, Belendez A, Pascual I. Generation of diffractive optical elements onto a photopolymer using a liquid crystal display. In: SPIE Proceedings 7717, Optical Modelling and Design, 77170D; 14 May 2010; Brussels, Belgium. SPIE; 2010. DOI: 10.1117/12.854786

[34] Goodman JW. Introduction to Fourier Optics. 2nd ed. New York: McGraw-Hill; 1987. 528 p

Chapter 9

The Effects of Casting and Blending on Properties of Ionomer and the Electromechanical Responses of Ionic Polymer Metal Composite Actuators

Yanjie Wang, Jie Ru, Denglin Zhu, Minzhou Luo,
Xiaofeng Liu and Hualing Chen

Additional information is available at the end of the chapter

http://dx.doi.org/10.5772/intechopen.72440

Abstract

As one typical kind of ionic electroactive polymers (iEAPs), ionic polymer metal composites (IPMC) consist of an ionomer and two thin layers of metallic electrode on its both sides. The micro-properties of the ionomer, usually Nafion as the most used ionomer, exert strongly effects on the responses of IPMC actuator. Our works revealed the effects of casting process with different additives (ethylene glycol (EG), dimethyl sulfoxide (DMSO), N, N′-dimethyl formamide (DMF) and N-methyl formamide (NMF)), and blending with sulfonated multi-walled carbon nanotube (sMWCNT) on properties of ionomer and the electromechanical responses of IPMC actuators. Some important properties of casting membrane and sMWCNT/Nafion blending membrane, such as surface morphology, water uptake and ionic exchange capacity, etc., were measured and evaluated. Among the casting membrane-based IPMC actuators, EG based IPMC actuator has larger deformation at 2 V DC voltage. And a trace amount of sMWCNT can improve the performances of IPMCs significantly for realistic applications.

Keywords: ionomer, IPMC actuator, casting, blending, electromechanical responses

1. Introduction

As a typical kind of copolymers, an ionomer is an ion containing polymer, which consists of nonionic repeat units and a small amount of ion containing repeat units [1]. In an ionomer, the nonpolar chains are grouped together and the polar ionic groups are attracted to each other. The attractions of ionic units result strongly influence the polymer properties. Due to the presence of non-ionic groups and ionic groups, the membranes made with ionomers are used to achieve special ion-exchange ability, such as letting water molecules pass through and not the metal ions. One specific ion-selective membrane is a perfluorosulfonate ionomer which is called Nafion by DuPont, and the chemical formula is shown in **Figure 1**. The microstructure of Nafion varies in the length of back bones (I) and branches (II) and in the nature of the ionic side group, usually sulfonate anions (III). Usually, Nafion resins can be made into films or tubes and used in the process of production of chlorine, regeneration of spent acids, separations in chemical processing, lithium-sulfur batteries, as well as used in fuel cells and electrodialysis. [2–5]

In past two decades, significant attention has been directed to the field of ionic electroactive polymers (iEAPs), which generates quick and large strain due to the mass transfer effects in itself. [7] As one typical kind of iEAPs, ionic polymer-metal composites (IPMC) are very suitable for actuators and sensors with many unique advantages of large deformation, low actuating voltage, high compliance, lightness, softness, etc [8–10]. A typical IPMC has a sandwich structure (**Figure 2(a)**), which consists of a matrix membrane, usually Nafion membrane, and two thin metallic electrode layers on both sides of the membrane. So far, Nafion remains as the benchmark for a majority of research and development in IPMC technology [11].

The actuation mechanism of IPMC involves a variety of physical and chemical processes. When an external voltage is applied at the both sides of IPMC, it will induce an electric field gradient across the thickness direction. The ions carrying some solvent molecules will be dragged from one side to another side and results in difference in ion concentrations across the thickness direction. Then the expansion and contraction on both sides of IPMC occur, which consequently results in bending deformation of IPMC (**Figure 2(b)**) [12].

One of the most important factors seriously affecting the performance of IPMC is the properties of Nafion, which mainly provides the backbone and ion transport medium of IPMC during the actuating process. Commercial Nafion series have a relatively poor thickness range from 50 to 250 μm, which will finally result in low output force. From the perspective of increasing the output force of IPMC, the easiest and most efficient way is to increase the thickness of the matrix membrane. So far, there are two ways to form thick Nafion: hot pressing [13] and solution casting [14], developed by Lee and Kim, respectively. The former uses hot-press technology to integrate multi-layers of commercial Nafion membranes, which easily result to delamination after actuating IPMC repeatedly [15]. The latter overcome this defect, which can obtain the ionomer with arbitrary shape and thickness. However, the work from [14] did not report more information about the micro-properties of casting ionomer. Other researchers [16–18] researched the formation of casting ionomer a lot for fuel cell and chlor-alkali application. They revealed that temperature, curving time, and the choice of additive affected

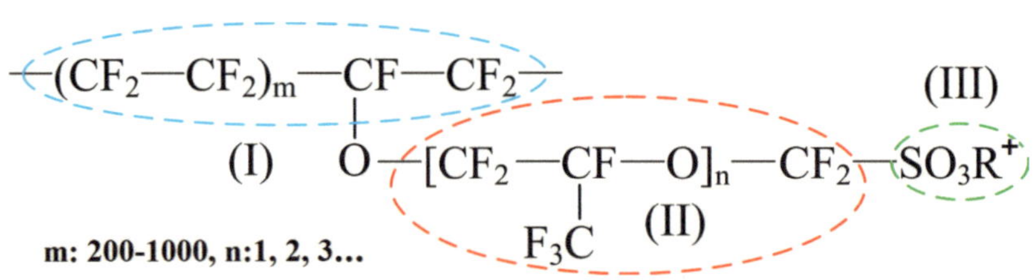

Figure 1. A typical chemical formula of Nafion series [6].

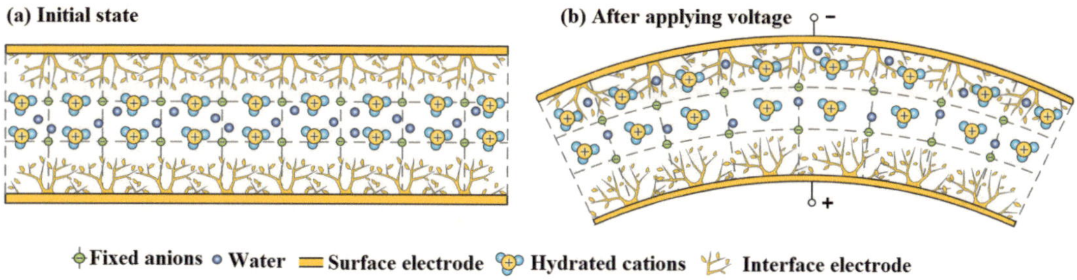

Figure 2. Schematic diagram of the configuration (b) and actuation mechanism of IPMC.

microstructure of ionomer [19, 20], in which the effects of curving time and temperature on the micro-properties of casting ionomer have been clearly clarified. From the perspective of the additives, Moore et al. [21] has confirmed that casting ionomer would display worse mechanical intensity and higher solubility without the addition of the high boiling additives. Then researchers [22–25] began to pay attention to the effect of additives and tried to find out that why and how additives seriously affect the morphology formation of casting ionomer. And they experimentally confirmed that the addition of the high boiling additive changed the water content, conductivity and modulus, of casting ionomer. These factors also will exert significant influences on the electromechanical responses of IPMCs based on the casting ionomer.

Another method to improve the performance of IPMC is to blend conductive or nonconductive compounds into Nafion matrix. The emerging nano-scale carbonaceous materials, such as carbon nanotubes (CNTs), graphene, fullerenes, carbon nano-fibers (CNF) and carbon-filled materials and so on, have prompted the research into a new class of actuators consisted of "smart nano-materials"-carbonaceous materials [26–29]. The mentioned materials just now typically have the advantages of remarkably large specific surface area together with nano-porous structure to guarantee a lot of insertion sites for electrolyte ions [30]. Due to the high strength, stiffness and excellent electrical conductivity of CNTs, CNTs-based composites are widely applied in various research fields [31–33]. CNTs are also applied to improve the performances of Nafion-based IPMCs. Sulfonated multi-walled carbon nanotube (sMWCNT), with sulfonic acid groups (-SO3H) covalently bonding to the surface of MWCNT and very weak Van der Waals' force among the bundles, is water-soluble and possess numerous insertion sites while the mechanical properties are hardly weakened compared to the original MWCNT [34–36]. Therefore, it has been widely used in super capacitors [37, 38] and lithium ion batteries [39, 40]. Considering the working mechanism and characteristics of the Nafion-based IPMCs, sMWCNT is expected to improve the mass transfer performance and elastic modulus, and then result in improvements on the electromechanical and electrochemical performances of the IPMC actuators.

2. Experimental section

2.1. Experimental materials

The 5 wt. % Nafion® dispersions (D520) were purchased from DuPont TM. Commercially available Nafion® 117 membrane (N117), purchased from DupontTM, was used as a reference.

Concentrated nitric acid (69.5 wt. %), sodium hydroxide Sodium p-aminobenzenesulfonate, $NaNO_2$, $Pd(NH_3)_4Cl_2$, NaBH4, and $N_2H_4 \cdot H_2O$ were purchased from Aldrich. Four additives for casting membranes were purchased from TianLi Chemical Co., including EG, DMSO, DMF and NMF. MWCNT was purchased from the Chinese Academy of Sciences Chengdu Organic Chemical Co. Ltd. All of the materials were analytically pure and used without further purification.

2.2. Preparation process

2.2.1. Solution casting process

We performed a typical solution casting process of Nafion ionomer as follows: Firstly, we separately pipetted 50 mL of 5 wt. % Nafion® dispersions into four glass vessels. Subsequently, we in sequence added EG, DMSO, DMF and NMF into four glass vessels with the fixed quantity of 5 mL respectively. And then treated the solutions by ultrasonic for 0.5 h for the purpose of uniformity. The solutions were put into oven under vacuum condition with a constant temperature of 80°C for 0.5 h. After the volatilization of solvents with low boiling point, the temperature of the oven was turned up to 120°C for further curing process. The curing time was set up for 6 h. After the oven cooled down, the glass vessels were took out and then immersed into cool water. The casting ionomer peeled off naturally 5 min later. To get the acid form of the ionomers, they were immersed in boiling 5% H_2O_2, 0.5 M H_2SO_4 solution and DI water for 1 h, respectively, and finally saved in DI water. As shown in **Figure 3**, the casting process mainly consists of four steps.

For the convenience, we mark the casting membranes with the additive of EG, DMSO, DMF and NMF as EG, DMSO, DMF and NMF, respectively. During the casting procedure, we only control additive as a variable. Other parameters, such as the precursors, mixed ratio and heat treatment, were employed from the literature [41], as illustrated in **Table 1**.

2.2.2. sMWCNT/Nafion blending membrane

A certain amount of sMWCNT was added to 3 mL ethylene glycol (EG) and sonicated for 30 min, then the suspensions was mixed with the Nafion solution and stirred for 1 hr. in a poly(dimethylsiloxane) (PDMS, Sylgad184) container (40 × 60 × 40 mm). The mixture was evaporated at 90 and 100°C for 12 h successively to cast sMWCNT/Nafion hybrid membrane.

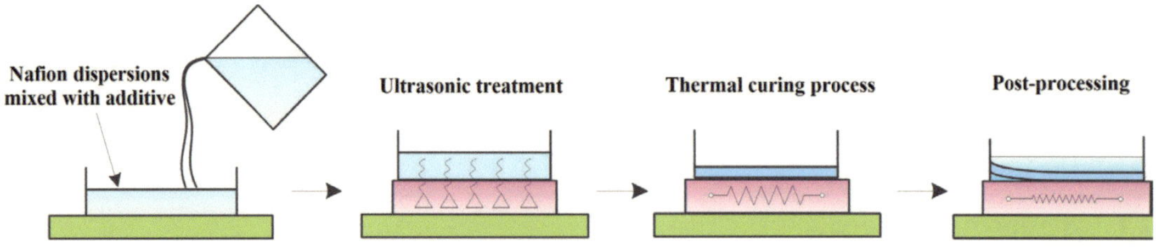

Figure 3. Solution casting procedure.

Samples	Nafion® precursor	Additives	Mixed ratio	Heat treatment (°C, h)
EG	D 520	EG	10:1	120, 6
DMSO	D 520	DMSO	10:1	120, 6
DMF	D 520	DMF	10:1	120, 6
NMF	D 520	NMF	10:1	120, 6
Nafion 117[a]	—	—	—	—

[a]data from Nafion 117 are tested as a contrast.

Table 1. Parameters set-up during casting process.

After that, the membrane was annealed at 120°C for 1 h. The hybrid membranes were obtained with sMWCNT weight fraction of 0, 0.25 and 0.5 wt%, respectively. Another hybrid membrane with MWCNT content of 0.5 wt% was also prepared for comparison. The sizes of Nafion membranes are 60 × 40 mm × (210 ± 10) μm (length × width × thickness) under dry state. The composition of the casting solutions and membranes are summarized in **Table 2**.

2.3. IPMC fabrication

The Pd-IPMCs were fabricated via assembling a hybrid Nafion layer and two palladium electrode layers by the impregnation-reduction method, which has been described in detail in our previous report [42]. The major steps were as follows: 1) Surface roughening treatment. The membranes were roughened with a sandblasting machine, then washed with 20% ethanol solution in an ultrasonic cleaning machine and boiled in 1 M HCl and water for 30 min, successively. 2) Ion adsorption. This step was to soak the membranes in $Pd(NH_3)_4Cl_2$ solution to adsorb $[Pd(NH_3)_4]^{2+}$ via an ion-exchange process. 3) Reduction. In this step, the adsorbed $[Pd(NH_3)_4]^{2+}$ was reduced to metallic state by strong reducing agent $NaBH_4$ to form infiltrate electrode. 4) Further plating. In order to increase the thickness of the surface electrode and reduce the surface resistivity effectively, $Pd(NH_3)_4Cl_2$ and H_2NNH_2 were put into the same solution simultaneously to grow Pd nano-particles above the infiltrate Pd electrode layer. 5) Ion exchange. This step was to exchange H^+ into Li^+ (the working ions) by soaking the IPMC strips in 2 M LiCl solution at room temperature for 24 h. The size of IPMC specimens for final characterization was 30 × 5 mm × (220 ± 10) μm (length × width × thickness) under hydrated condition.

Membrane type	Casting solution composition			Membrane composition	
	CNT (mg)	Nafion (aq) (g)	EG (g)	CNT (mg)	Nafion (s) (g)
Pure Nafion	0	19.00	4.5	0	0.9500
0.25 wt% sMWCNT	2.375	18.95	4.5	2.4	0.9475
0.5 wt% sMWCNT	4.750	18.90	4.5	4.8	0.9450
0.5 wt% MWCNT	4.750	18.90	4.5	4.8	0.9450

Table 2. Composition of the casting solutions and membranes.

2.4. Characterizations

2.4.1. Characterization of ionomer morphology

We examined the microstructure properties of the casting ionomers using AFM equipment from BRUER, USA with resonance frequency of 380 kHz and selection of tapping mode, which was performed at room temperature and humidity. Due to the fact that surface nanomorphology of the casting ionomers was affected easily by the change of room humidity, we mounted the samples on a silicon substrate. The surface and cross-section morphology of so based IPMCs were observed by SEM (VG3210677 and Hitachi SU-8010). Scanning electron microscopy (SEM) was performed at an accelerating voltage of 20.0 kV. All sample cross-sections were obtained by low-temperature cracking, being placed in liquid nitrogen for 5 min and then broken into pieces.

2.4.2. Characterizations of the membranes

2.4.2.1. Differential scanning calorimetry (DSC) and thermogravimetric (TG) measurements

We employed a Thermal Analysis Model DSC1 from Mettler Toledo, Switzerland to perform DSC analysis of the ionomers and so based IPMCs. The weights of test samples were at the range of 4–8 mg. Before DSC observations, the surface of each sample needed to be treated in an oven with the temperature of 80°C for 2 h. The heating rate 10°C/min and N_2 flow rate 80 ml/min of DSC was selected. Thermogravimetric (TG) measurements of sMWCNT and MWCNT were carried out by using a TG analyzer (STA 449C, NETZSCH, Germany) at a heating rate of 20°C min^{-1} from 20 to 800°C and nitrogen flow rate of 30 mL min^{-1}.

2.4.2.2. Water-uptake ratio (WUR)

In order to measure the WUR, the membranes with different sMWCNT weight fraction were soaked in DI water for 24 h to be fully saturated; then the membranes were weighed immediately after wiping the water on the surface with filter paper, and the mass was recorded as M_{wet} (g). After that, the hydrated membranes were dried at 100°C for at least 24 h until a constant mass was obtained, which was recorded as M_{dry} (g). The WUR (w) of each membrane can be calculated according to Eq. (1).

$$w(\%) = \frac{M_{wet} - M_{dry}}{M_{dry}} \times 100 \tag{1}$$

2.4.2.3. Ionic exchange capacity (IEC)

The IEC was measured using the acid–base titration method with KOH (aq) and HCl (aq) described in detail in ref. [36] and calculated according to Eq. (2).

$$IEC = \frac{V_{KOH} \times N_{KOH}}{m_{membrane}} (meq/g) \tag{2}$$

where V_{KOH} is the quantity (ml) of KOH (aq), N_{KOH} is the normality (mol) of KOH (aq) and membrane is the mass of the membrane. A minimum of three sets of experiments were performed until the percentage error was well within the experimental limits.

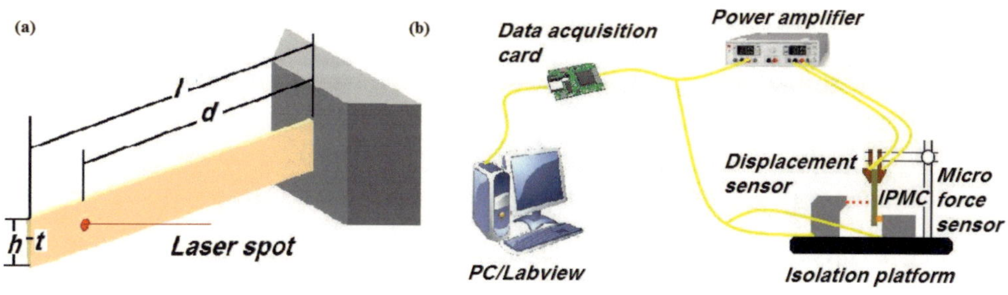

Figure 4. Schematic illustration for dimensions of IPMC strip (a) and experimental apparatus(b).

2.4.3. Electromechanical test of IPMCs

The equivalent elastic modulus of IPMCs was measured using the free oscillation attenuation method, and the geometry of the IPMC specimens for the stiffness and electromechanical tests in this work is illustrated in **Figure 4(a)**. During the test, the specimens were prescribed a small initial deformation firstly, and then the natural frequency f of the specimens was obtained by a fast Fourier transform of the free vibration response curve. Every specimen was tested in wet state in two minutes. The equivalent elastic modulus can be calculated via Eq. (4), which is derived from Euler-Bernoulli beam theory.

$$E = \left(\frac{2\pi}{3.52}\right)^2 \frac{m f^2 l^3}{I} = 3.87 \frac{\pi^2 m f^2 l^3}{h t^3} \quad (3)$$

Electro-active performance was characterized by measuring the displacement and blocking force of IPMC specimens using the same test apparatus described in detail in ref. [42]. The specimen was clamped by a gold-plated copper clamp on one end firstly, while the other end is freestanding with a free length of 30 mm. Then the specimen was applied a constant potential by a Labview software and an arbitrary power supply (HM8143). The displacement and blocking force were measured with a laser displacement sensor (Keyence LK-G80) and a micro force transducer (Transducer Techniques, GSO-10) at the measuring distance of 15 mm, respectively. All the measurements were carried out in air. The current, potential, displacement and blocking force were recorded by the Lab view software simultaneously for 300 s. The experimental apparatus is shown in **Figure 4(b)**.

3. Results and discussions

3.1. Morphology analysis

As shown in **Figure 5**, we obtain AFM topography micrographs of polymer chain conformations of the sample EG (a), DMSO (b), DMF (c), NMF (d), with Nafion 117 (e) as a reference. In each subfigure of **Figure 5**, we defined that bright region is high both in topography and in phase, in which high phase represents the backbone consisted of long polymer chains under tapping mode. Likewise, the gray regions in height reveal the ionic or nano-porous domains with a diameter range of 5–50 nm. We can see some significantly differences by

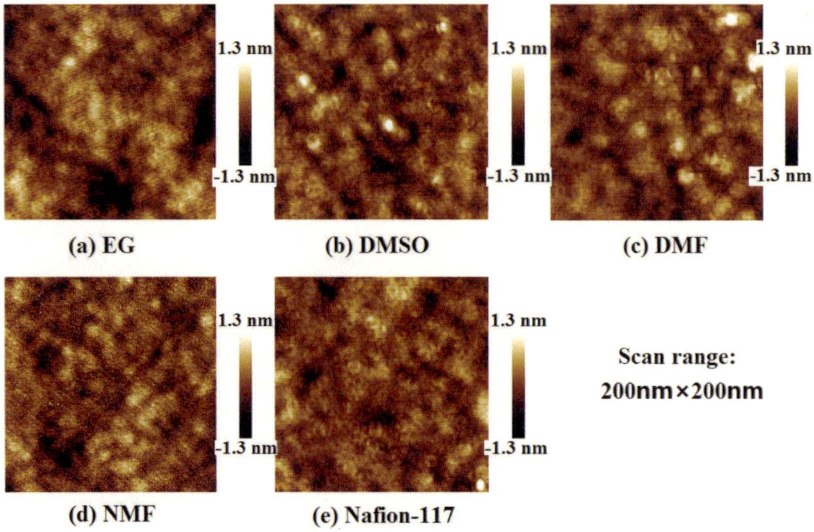

Figure 5. AFM topography images (200 × 200 nm). Images (a), (b), (c), (d), and (e) correspond to the topography of EG, DMSO, DMF, NMF and Nafion 117, respectively.

these subfigures of **Figure 5**. It is interesting that backbones conformations of casting ionomers are more amorphous and loose in **Figures 5(a)** and **5(b)**. On the other side, the polymer chains together with ionic domains are more uniform and compact in **Figure 5** (c), (d) and (e), which is responsible for the differences of the compatibilities between Nafion chain molecules and additives. When evaporated from casting solutions, additives assist Nafion side chain sulfonic acid groups to move along with perfluorocarbon backbones so as to form solid membranes. Meanwhile, backbone aggregations in ionomers result in the formation of more sulfonic acid group aggregations, i.e. larger size of ionic clusters, which had been revealed by Ma et al. [22].

Figure 6 shows the inner morphology of P-membrane (a), M-membrane (b, d) and S2-membrane (c, e) at different magnification levels. The two hybrid membranes presented quite rough surfaces, while the P-membrane presented a typically smooth polymer surface with several wrinkles. For the S2-membrane, sMWCNT was homogeneously dispersed in Nafion matrix without any entangled structure or obvious agglomeration, which is advantageous to improve the electrochemical and electromechanical performances of the reinforced membranes and corresponding IPMCs. It is also obvious that the dispersion of sMWCNT was much more uniform than that of MWCNT in Nafion matrix, in that the hydrophilic nature of sMWCNT facilitated their dispersion in the Nafion matrix via the hydrophilic interaction between -SO_3H groups of Nafion and sMWCNT [43]. The electrode morphology of the S2-IPMC is also shown in **Figure 8f**. The thickness of the Pt layer on the membrane surface of each IPMC was around 11–16 μm, and the range of the surface resistance of each IPMC was around 2–5 Ω/□. As a result, the surface electrodes, without remarkable difference, had no direct correlation with the increasing content of the sMWCNT. Systematic and quantitative correlations of the surface electrode with other fundamental parameters are complex, which is beyond this work and will be discussed in our forthcoming works.

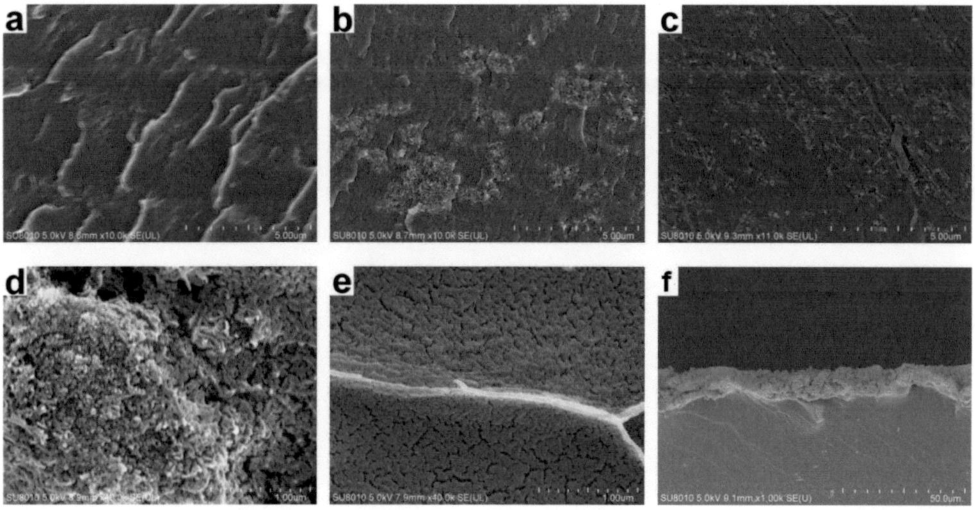

Figure 6. The inner morphology of P-membrane (a), M-membrane (b, d) and S2-membrane (c, e) at different magnification levels, and the electrode morphology of the S2-IPMC (f).

3.2. DSC and TG analysis of casting and blending Nafion

In order to evaluate the influence of the additives on thermal behavior of the casting membranes and IPMCs, DSC was performed from 0 to 300°C. We divided the overall morphology change of the sample into two temperature points based on the temperature range. 110°C is the first temperature point (Tg), which appears an intense endothermic valley. We thought the reason was due to the glass transition of the polymeric matrix, and the other temperature point ™ is about 230°C [44]. The enthalpy change displayed crystalline region is very little inside the casting ionomers. By testing, we got the Tgs of the samples, which were 104.90, 131.37, 114.81, 109.39 and 117.58°C in sequence. Their Tg values were similar, during which the Tg of DMSO was highest. The difference of Tgs may be attributed to the morphology change of the polymer chain structure and ion clusters [45]. From the DSC results, the Tms of the samples are almost the same, which only exhibit a small difference. We considered that the curing temperature would result to form the same percentage of crystalline region inside the ionomer. DSC results of casting ionomers are shown in **Figure 7** and relative parameters are summarized in **Table 3**.

From **Figure 8a**, it can be found that the weight of SCNT nearly decreased by 10%, while the weight of MWCNT only decreased by 5%. In the curve of MWCNT, the small weight loss around 100°C was caused by the evaporation of absorbed water molecules, while the weight loss between 220 and 400°C was most likely due to the elimination of amorphous carbon [46]. In comparison with MWCNT, the sMWCNT showed a slightly prominent weight loss, which was mainly caused by the decomposition of the grafted azobenzene-4-sulfonic acid and the elimination of amorphous carbon. The slow weight loss between 30 and 130°C was also due to the evaporation of absorbed water molecules. The weight decreased dramatically in the range from 220 to 800°C, which was caused by the decomposition of the side chains of sMWCNT and the elimination of amorphous carbon [40]. The results can also indicate that the sMWCNT would be thermally stable in the membrane casting process.

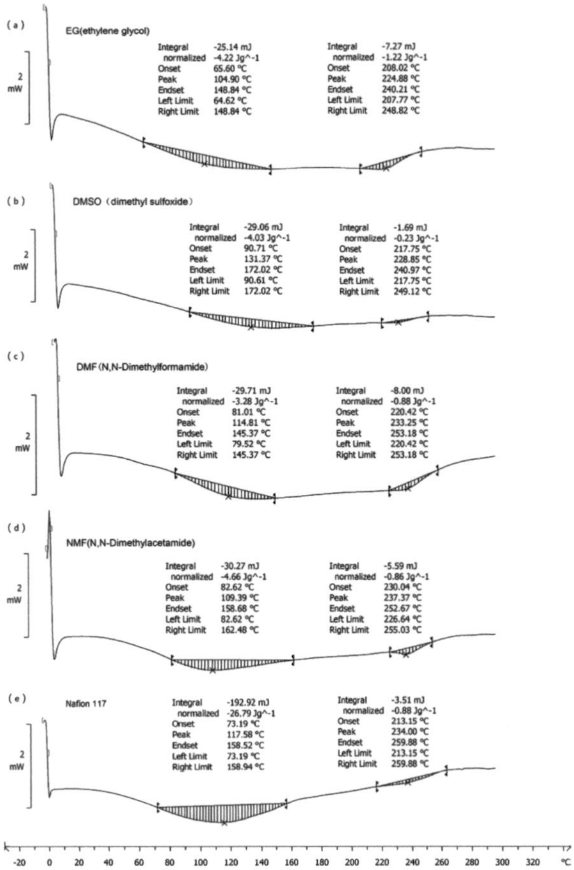

Figure 7. The DSC results of sample EG(a), DMSO(b), DMF(c), NMF(d) and Nafion 117(e).

Materials	Tg/°C	ΔH/mJ	Tm/°C	ΔH/mJ
EG	104.90 ± 2.1	−25.14	224.88 ± 2.4	−7.27
DMSO	131.37 ± 1.6	−29.06	228.85 ± 3.2	−1.69
DMF	114.81 ± 2.8	−29.71	233.25 ± 4.2	−8.00
NMF	109.39 ± 2.2	−30.27	237.37 ± 3.2	−5.59
Nafion 117	117.58 ± 3.1	−192.92	234.00 ± 2.6	−3.51

Table 3. Thermal analysis (*Tg,Tm*) of Nafion, modified Nafion and so-based IPMCs.

The TG curves of the S2-membrane and M-membrane shown in **Figure 8b**, with the same trend, indicate that the hybrid membranes are thermally stable up to ca. 300°C, which is the same as that of the P-membrane. The similar curves were also obtained by Lage [47] and Almeida [48], respectively. The gradual weight loss in the range between 30 and 300°C was mainly due to the evaporation of water molecules. The thermal degradation of the membranes occurred in three stages: the initial stage between 300 and 400°C was due to the desulfonation process of -SO_3H

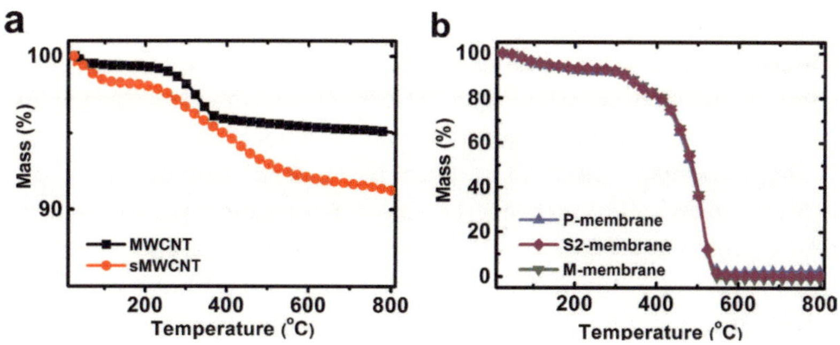

Figure 8. The TG analysis of MWCNT and sMWCNT(a), the P-membrane, S2-membrane and M-membrane(b).

groups, while the second stage between 400 and 480°C was caused by the decomposition of the branch chain from the polymer backbone, and the final stage between 480 and 550°C was related to the decomposition of PTFE (polytetrafluoroethylene) chain. It can be concluded that the hybrid membranes had the same thermal stability as that of the P-membrane.

3.3. WUR and IEC of the casting and blending Nafion

According to vehicle mechanisms and Grotthus 'hopping' theory, water-uptake ratio (WUR) and ionic exchange capacity (IEC) are important factors that strongly affect the hydrated cation migration in Nafion membranes and in terms of mechanical strength and actuation performance of the resulting IPMCs [49, 50]. The water contents of the casting membranes are summarized in **Figure 9**. Evidently, EG and DMSO have higher water contents than that of the Nafion 117 in fully hydrated states. The water contents follow the sequence: EG > DMSO > Nafion 117 > DMF > NMF.

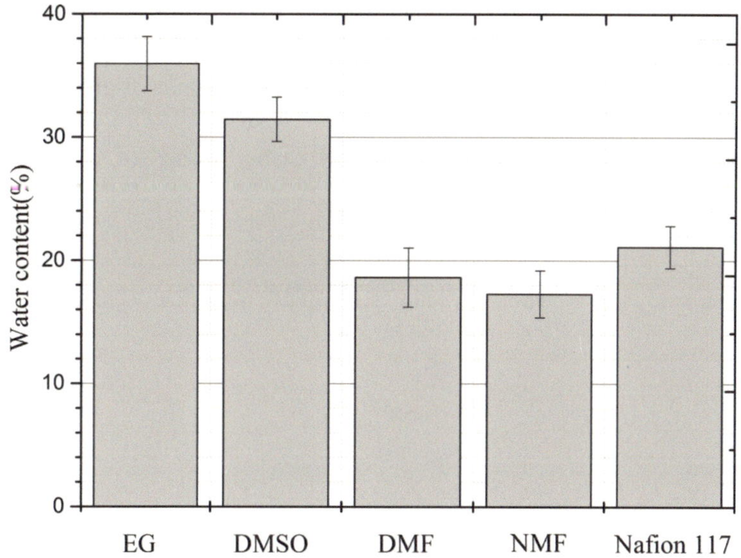

Figure 9. Water content of the samples.

These results show that the additives exert an important effect on the conformational formation. When the ionomer was absorbed enough water molecules, the conformation of ionic clusters will present different morphologies. The conformation formation of ionomer is very complex due to the interaction between polymer backbone and ionic clusters. When solvents were heated and evaporated out of casting solutions, the side chains of ionomer will curve and shrink to form solid ionomers. During this process, the ionic aggregation occurs as well. Generally, the water content of ionomer is proportional to the size of ionic aggregations [45]. From **Figure 8**, the results confirmed that the additive exerted its impact on the water content of the casting ionomer, which was similar to the results reported by the reference [22].

The IEC of ionomer decides by the quantity of ionic groups in fixed volume of ionomer, which is consistent to the number of H^+ in the ionomer. Usually, the increase of the number of sulfonate will result to the increase of IPMCs performance. **Figure 4** shows IECs of samples. As shown in **Table 4**, the used additives lead to the differences of the IECs, of which the IEC of Nafion 117 is highest.

The WUR and IEC of the membranes with different CNT content are summarized in **Table 5**. The WUR increased as the content of sMWCNT in the polymer matrix increased. The WUR of the S2-membrane was nearly increased by 20% with comparison to that of the P-membrane. On the contrary, there was a sharp decrease in WUR of the M-membrane (decreased by 15%), which is consistent with other Nafion-nano-filler systems like MWNT-Nafion [51] and graphene-Nafion [33]. It has been proposed that MWCNT is a hydrophobic particle [51], the substitution of Nafion by MWCNT would lead to a dramatical decrease in water absorption ability of the membrane. Even so, great changes in its hydrophobicity would occur [30] when MWCNT is subjected to chemical modification of MWCNT, like grafting and coating. In this research, the hydrophilic -SO_3H groups were grafted onto the surface of MWCNT, which provided sMWCNT with the sites of hydrogen bonding with water. Therefore, there was a significant increase in the water absorption capacity of all the sMWCNT/Nafion membranes than that of the M-membrane. It can also be seen that the IEC increased slightly with the content of sMWCNT increasing in the polymer matrix, which is attributed to the great proton exchange ability of -SO_3H groups. In addition, the increase in WUR and IEC of the sMWCNT/Nafion membranes maybe due to the sMWCNT increasing the size of the ion clusters and the number of exchange sites for each cluster [52].

Samples	**Ion-exchange capacity (meq/g)**
EG	953 ± 14
DMSO	978 ± 22
DMF	996 ± 10
NMF	941 ± 11
Nafion 117	1104 ± 16

Table 4. Ion-exchange capacity of ionomer samples.

CNT content	WUR (%)	IEC(meq/g)
0%	22.74	0.79
0.25% sMWCNT	23.54	0.82
0.5% sMWCNT	27.06	0.93
0.5% MWCNT	19.44	0.84

Table 5. Properties of the membranes.

3.4. Electromechanical properties evaluation of IPMCs

Figure 10 represents the deformations of the EG, DMSO, DMF, NMF, and Nafion 117-based IPMC actuators as a function of time at 2.0 V DC voltages, the tip displacements of which show 4.02, 3.22, 3.17, 1.1 and 6.9 mm, respectively. Nafion-based IPMC exhibits a highest tip displacement. And EG-based IPMC has a larger tip displacement than other samples. This reason is attributed to the higher water content of EG-based IPMC than other samples.

In order to study the dynamic response, the change histories of the deformation of the IPMCs under sinusoidal stimulation under the voltage amplitude of 1 and 2 V at an excitation frequency of 0.1 Hz were recorded and shown in **Figure 11a** and **b**, respectively. These results indicate that the deformation of the IPMCs can be controlled well using low sinusoidal voltages. Notably, the peak-to-peak (P–P) displacement increased significantly with respect to the increasing content of sMWCNT. For the S2-IPMC, the maximum P–P displacements were

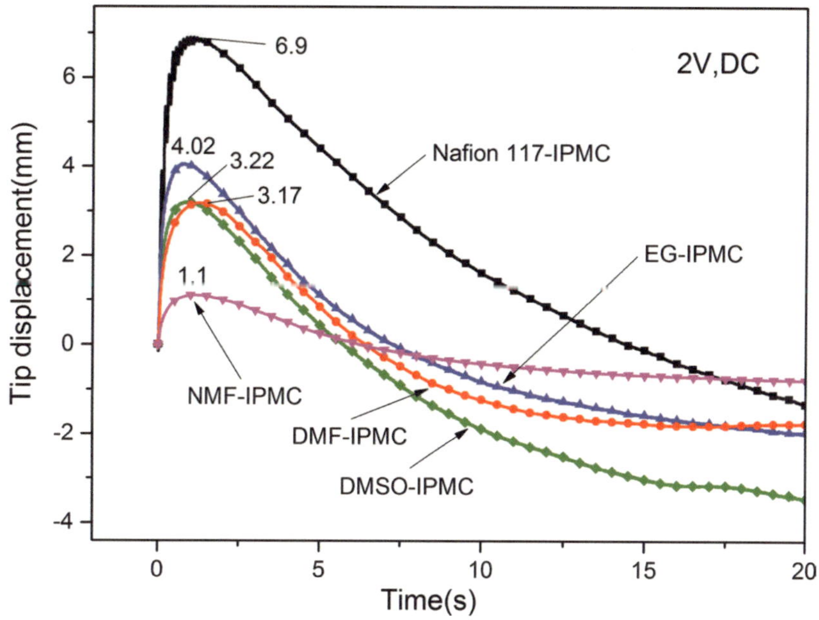

Figure 10. The deformations of the EG, DMSO, DMF, NMF, and Nafion 117-based IPMC actuators.

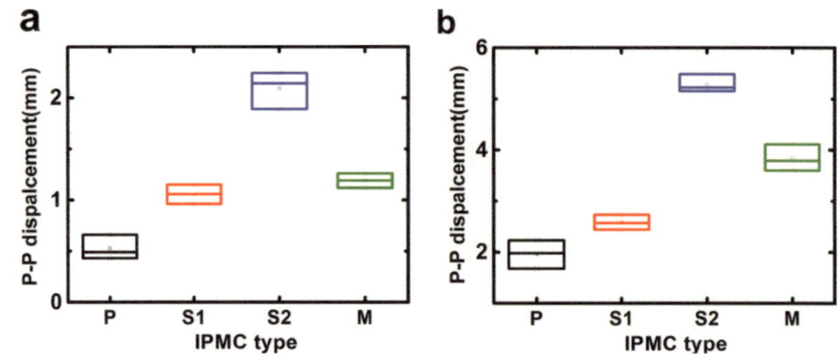

Figure 11. The peak-to-peak (P–P) displacement of harmonic response of the IPMCs under AC 1 V (a) and AC 2 V (b) with the excitation frequency of 0.1 Hz (P: P-IPMC, S1: S1-IPMC, S2: S2-IPMC, M: M-IPMC).

measured to be about 2.06 and 5.28 mm at the driving voltages of 1 and 2.0 V, respectively. In contrast, the maximum P–P displacements of the P-IPMC were 0.40 and 1.96 mm, and the maximum P–P displacements of the M-IPMC were 1.19 and 3.83 mm at the same driving voltages, respectively. The DC excitation measurements have also been carried out under DC 1 and 2 V at 0.02 Hz, respectively. The results, with the same trend to that of the dynamic response, are shown in **Figure 12**. The maximum displacement of the IPMCs also increased with respect to the content of sMWCNT in the Nafion matrix.

It can be demonstrated that the sMWCNT reinforced IPMCs have much more excellent bending performance than P-IPMC and M-IPMC. Due to the favorable hydrophility, ion-exchange capacity and proton conductivity, sMWCNT showed a higher degree of solvated cation diffusion with comparison to MWCNT. With the incorporation of sMWCNT into Nafion matrix, the improvement in potential difference between the interface electrodes could promote the flux of cations. As a result, the incorporation of sMWCNT induced significant improvement in the deformation of the resulting IPMCs.

Figure 13 shows the maximum peak-to-peak displacement versus the excitation frequency of the harmonic responses under AC 2 V. Obviously, the maximum P–P displacements decreased remarkably as the excitation frequency increased, indicating an inverse relationship between the deformation and the excitation frequency. The decrease in deformation is

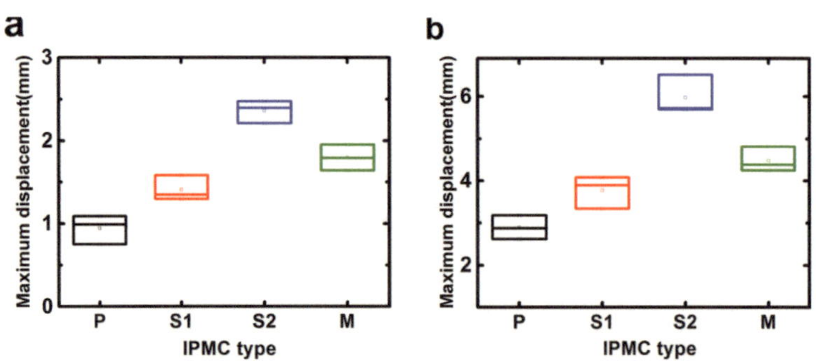

Figure 12. The maximum displacement of step response of the IPMCs under DC 1 V (a) and DC 2 V (b) with the excitation frequency of 0.02 Hz.

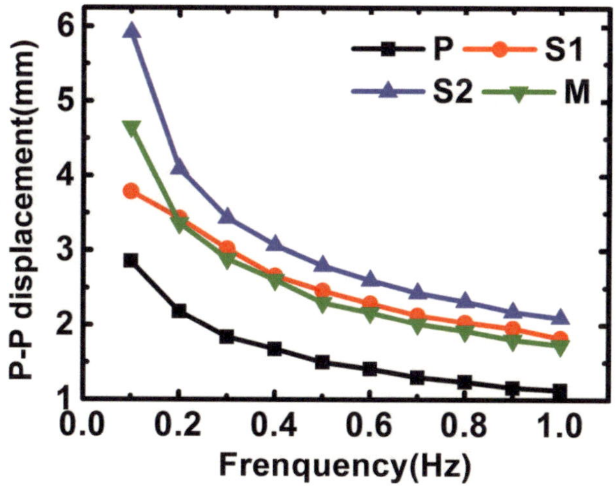

Figure 13. The maximum peak-to-peak displacement versus the excitation frequency of the harmonic responses under AC 2 V.

caused by shortening the driving time of the potential as the frequency increased. Even so, the harmonic response of the S2-IPMC is much better than those of the P-IPMC and M-IPMC, and that of the SWCNT-Nafion IPMCs with the content of 14 wt% [53]. These results can also demonstrate that a trace amount of the sMWCNT can significantly improve the harmonic response and bending deformation of IPMCs. This may be due to that the reinforcing filler-sMWCNT can effectively shorten the hopping distance and reduce the negative impact of water diffusion [54].

4. Summary and conclusions

The effects of casting and blending on the performances of ionomers were analyzed, using a series of experimental tests, as well as the so based IPMCs. The ionomers were successfully obtained based on commercial Nafion® dispersion by solution casting process. The sMW-CNT/Nafion membranes and crossponding IPMCs were prepared and characterized as well. The morphologies of the membranes were characterized by AFM topography. Among the casting membrane-based IPMC actuators, EG based IPMC actuator has larger deformation at 2 V DC voltage, whose electromechanical property is most close to that based on Nafion 117. And it is clearly that EG is a more preferable additive during the casting process for the enhancement of IPMC performance. Due to the improvement of water-uptake ratio, proton conductivity and elastic modulus of sMWCNT/Nafion blending membrane, a superior bending deformation and carrying capacity were observed in the sMWCNT/Nafion IPMCs.

Our work has confirmed that additives have great influences on the mechanical properties of the casting membranes and consequently electromechanical coupling of IPMCs. And the sMWCNT/Nafion blending membrane would be promising candidates for use in IPMC actuators. Further research works will focus on exploring more methods to improve the performance of the so based IPMCs for different application background.

Acknowledgements

This work is supported by the National Natural Science Foundation of China (NO.51505369 and 51375140), the key development program of Jiangsu Province in China (NO.BE2017071), the Foundation (NO.M20133004) of Changzhou Key Laboratory of Special Robot and Intelligent Technology, the Project of International Cooperation and Exchange of Changzhou (NO. BE2017071) and the Fundamental Research Funds for the Central Universities, P.R. China. The authors gratefully acknowledge the supports.

Author details

Yanjie Wang[1†*], Jie Ru[2†], Denglin Zhu[1], Minzhou Luo[1], Xiaofeng Liu[1] and Hualing Chen[2]

*Address all correspondence to: yj.wang1985@gmail.com

1 Department of Mechanical and Electrical Engineering, Hohai University, Changzhou Campus, Changzhou, People's Republic of China

2 Department of Mechanical Engineering, Xi'an Jiaotong University, Xi'an, People's Republic of China

†The first two authors contributed equally to this work.

References

[1] Zhang L, Brostowitz NR, Cavicchi KA, et al. Perspective: Ionomer research and applications. Macromolecular Reaction Engineering. 2014;**8**(2):81-99

[2] Naterer GF, Suppiah S, Stolberg L, et al. Progress of international program on hydrogen production with the copper-chlorine cycle. International Journal of Hydrogen Energy. 2014;**39**(6):2431-2445

[3] Bruno JE, Dooley KM. Regeneration of a supported Nafion® catalyst for the double-bond isomerization of octadecenes. Applied Catalysis A: General. 2016;**526**:70-76

[4] Bauer I, Thieme S, Brückner J, et al. Reduced polysulfide shuttle in lithium-sulfur batteries using Nafion-based separators. Journal of Power Sources. 2014;**251**:417-422

[5] Lopez-Haro M, Guétaz L, Printemps T, et al. Three-dimensional analysis of Nafion layers in fuel cell electrodes. Nature Communications. 2014;**5**:5229

[6] Rollet A L, Diat O, Gebel G. A new insight into Nafion structure. The Journal of Physical Chemistry B. 2002;**106**(12):3033-3036

[7] Tiwari R, Garcia E. The state of understanding of ionic polymer metal composite architecture: A review. Smart Materials and Structures. 2011;**20**(8):083001

[8] He Q, Yang X, Wang Z, et al. Advanced electro-active dry adhesive actuated by an artificial muscle constructed from an ionic polymer metal composite reinforced with nitrogen-doped carbon nanocages. Journal of Bionic Engineering. 2017;**14**(3):567-578

[9] Shen Q, Wang T, Kim K J. A biomimetic underwater vehicle actuated by waves with ionic polymer-metal composite soft sensors. Bioinspiration & Biomimetics. 2015;**10**(5):055007

[10] Wang Y, Chen H, Liu J, et al. Aided manufacturing techniques and applications in optics and manipulation for ionic polymer-metal composites as soft sensors and actuators. Journal of Polymer Engineering. 2015;**35**(7):611-626

[11] Jo C, Pugal D, Oh IK, et al. Recent advances in ionic polymer-metal composite actuators and their modeling and applications. Progress in Polymer Science. 2013;**38**(7):1037-1066

[12] Wang Y, Zhu Z, Chen H, et al. Effects of preparation steps on the physical parameters and electromechanical properties of IPMC actuators. Smart Materials and Structures. 2014;**23**(12):125015

[13] Lee SJ, Han MJ, Kim SJ, et al. A new fabrication method for IPMC actuators and application to artificial fingers. Smart Materials and Structures. 2006;**15**(5):1217

[14] Kim KJ, Shahinpoor M. A novel method of manufacturing three-dimensional ionic polymer-metal composites (IPMCs) biomimetic sensors, actuators and artificial muscles. Polymer. 2002;**43**(3):797-802

[15] He H, Zhan X, Wang L, et al. The preparation and electromechanical characteristic of the multilayer artificial muscle. Function Materials. 2011;**42**(B06):529-532

[16] Arimura T, Ostrovskii D, Okada T, et al. The effect of additives on the ionic conductivity performances of perfluoroalkyl sulfonated ionomer membranes. Solid State Ionics 1999;**118**(1):1-10

[17] Laporta M, Pegoraro M, Zanderighi L. Recast Nafion-117 thin film from water solution. Macromolecular Materials and Engineering. 2000;**282**(1):22-29

[18] Thomas TJ, Ponnusamy KE, Chang NM, et al. Effects of annealing on mixture-cast membranes of Nafion® and quaternary ammonium bromide salts. Journal of Membrane Science. 2003;**213**(1):55-66

[19] Li C, Sun G, Ren S, et al. Casting Nafion–sulfonated organosilica nano-composite membranes used in direct methanol fuel cells. Journal of Membrane Science. 2006;**272**(1):50-57

[20] Mohammadi F, Rabiee A. Solution casting, characterization, and performance evaluation of perfluorosulfonic sodium type membranes for chlor-alkali application. Journal of Applied Polymer Science. 2011;**120**(6):3469-3476

[21] Moore RB, Martin CR. Procedure for preparing solution-cast perfluorosulfonate ionomer films and membranes. Analytical Chemistry. 1986;**58**(12):2569-2570

[22] Ma CH, Yu TL, Lin HL, et al. Morphology and properties of Nafion membranes prepared by solution casting. Polymer. 2009;**50**(7):1764-1777

[23] Lin HL, Yu TL, Huang CH, et al. Morphology study of Nafion membranes prepared by solutions casting. Journal of Polymer Science Part B: Polymer Physics. 2005;**43**(21): 3044-3057

[24] Lee SJ, Yu TL, Lin HL, et al. Solution properties of nafion in methanol/water mixture solvent. Polymer. 2004;**45**(8):2853-2862

[25] Wang Y, Zhu Z, Chen H, et al. Effects of preparation steps on the physical parameters and electromechanical properties of IPMC actuators. Smart Materials and Structures. 2014;**23**(12):125015

[26] Fennimore AM et al. Rotational actuators based on carbon nanotubes. Nature. 2003;**424**: 408-410

[27] Guan W et al. Graphitic carbon nitride nanosheet electrode-based high-performance ionic actuator. Nature Communications. 2015;**6**:7258

[28] Woosung Y et al. Carbon nanotube-graphene composite for ionic polymer actuators. Smart Materials and Structures. 2012;**21**:055012 (7pp)

[29] Ru J, Wang Y, et al. Preparation and characterization of water-soluble carbon nanotube reinforced Nafion membranes and so-based ionic polymer metal composite actuators. Smart [29] Materials and Structures. 2016;**25**(9):095006

[30] Baughman RH et al. Carbon nanotube actuators. Science. 1999;**284**:1340-1344

[31] Lu L, Chen W. Biocompatible composite actuator: A supramolecular structure consisting of the biopolymer chitosan, carbon nanotube, and an ionic liquid. Advanced Materials. 2010;**22**:3745-3748

[32] Park IS, Kim SM, Kim KJ. Mechanical and thermal behavior of ionic polymer-metal composites: Effects of electroded metals. Smart Materials and Structures. 2007;**16**:1090-1097

[33] Thomassin JM et al. Beneficial effect of carbon nanotubes on the performances of Nafion membranes in fuel cell applications. Journal of Membrane Science. 2007;**303**:252-257

[34] Chang ZY et al. Hierarchically structured carbon-based composites: Design, synthesis and their application in electrochemical capacitors. Nanoscale. 2011;**3**:529-545

[35] Gupta V, Miura N. Influence of the microstructure on the supercapacitive behavior of polyaniline/single-wall carbon nanotube composites. Journal of Power Sources. 2006;**157**: 616-620

[36] Ham HT et al. Singlewall carbon nanotubes covered with polypyrrole nanoparticles by the miniemulsion polymerization. Polymer. 2005;**46**:6308-6315

[37] Hao Z et al. Tube-covering-tube nanostructured polyaniline/carbon nanotube array composite electrode with high capacitance and superior rate performance as well as good cycling stability. Electrochemical Communication. 2008;**10**:1056-1059

[38] Bo G et al. Preparation and electrochemical properties of polyaniline doped with benzene-sulfonic functionalized multi-walled carbon nanotubes. Electrochimica Acta. 2010;**55**: 2311-2318

[39] Seung IL C and Sang B L. Fast Electrochemistry of Conductive Polymer Nanotubes: Synthesis, Mechanism, and Application. Accounts of Chemical Research 2008;**41**:699-707

[40] Du CY et al. Sulfonation of carbon-nanotube supported platinum catalysts for polymer electrolyte fuel cells. Journal of Power Sources. 2008;**176**:9-15

[41] Silva RF, De M, Francesco A. Pozio. Electrochimica Acta. 2004;**49**:3211-3219

[42] Chang LF et al. Manufacturing process and electrode properties of palladium-electroded ionic polymer-metal composite. Smart Materials and Structures. 2012;**21**:065018

[43] Mahendran Rajagopalan and Il-Kwon Oh. Fullerenol-based electroactive artificial muscles utilizing biocompatible polyetherimide. ACS Nano. 2011;**5**:2248-2256

[44] R. Linacero, M. L. Rojas-Cervantes, J. De, D. Lopez-Gonzalez. Journal of Materials Science. 2000;**35**:3279-3287

[45] T. D. Gierke, G. E. Munn, F. C. Wilson. Polymer physics. Journal of Polymer Science. 1981;**19**: 1687-1704

[46] Chen CM et al. Purification of multi-walled carbon nanotubes by microwave digestion method. Diamond and Related Materials. 2004;**13**:1182-1186

[47] Lage LG et al. Thermal stability and decomposition of Nafion membranes with different cations using high-resolution thermogravimetry. Journal of Thermal Analysis and Calorimetry. 2004;**75**:521-530

[48] S H de Almeida et al. Thermal behavior of Nafion membranes. Journal of Thermal Analysis and Calorimetry. 1999;**58**:569-577

[49] Wang YJ et al. Effect of dehydration on the mechanical and physicochemical properties of gold- and palladiumionomeric polymer-metal composite (IPMC) actuators. Electrochimica Acta. 2014;**129**:450-458

[50] J Park et al. Electromechanical performance and other characteristics of IPMCs fabricated with various commercially available ion exchange membranes. Smart Materials and Structures. 2014;**23**:074001 (9pp)

[51] Jung J H et al. Electro active graphene-Nafion actuators. Carbon. 2011;**49**:1279-1289

[52] Bauer F, Pollard MW. Microstructural characterization of Zr-phosphate-Nafion- membranes for direct methanol fuel cell (DMFC) applications. Journal of Membrane Science. 2004:233-239

[53] Landi BJ et al. Single wall carbon nanotube-Nafion composite actuators. Nano Letter. 2002;**2**(11):1329-1332

[54] Paquette JW, Kim KJ. Ionomeric electroactive polymer artificial muscle for naval applications. IEEE Journal of Oceanic Engineering. 2004;**29**(3):729-737